T0140292

# Studies in Computational Intelligence

Volume 568

**Series editor**

Janusz Kacprzyk, Polish Academy of Sciences, Warsaw, Poland
e-mail: kacprzyk@ibspan.waw.pl

*About this Series*

The series "Studies in Computational Intelligence" (SCI) publishes new developments and advances in the various areas of computational intelligence—quickly and with a high quality. The intent is to cover the theory, applications, and design methods of computational intelligence, as embedded in the fields of engineering, computer science, physics and life sciences, as well as the methodologies behind them. The series contains monographs, lecture notes and edited volumes in computational intelligence spanning the areas of neural networks, connectionist systems, genetic algorithms, evolutionary computation, artificial intelligence, cellular automata, self-organizing systems, soft computing, fuzzy systems, and hybrid intelligent systems. Of particular value to both the contributors and the readership are the short publication timeframe and the world-wide distribution, which enable both wide and rapid dissemination of research output.

More information about this series at http://www.springer.com/series/7092

Paul Keng-Chieh Wang

# Visibility-based Optimal Path and Motion Planning

 Springer

Paul Keng-Chieh Wang
Department of Electrical Engineering
University of California
Los Angeles, CA
USA

ISSN 1860-949X
ISBN 978-3-319-35682-2
DOI 10.1007/978-3-319-09779-4

ISSN 1860-9503 (electronic)
ISBN 978-3-319-09779-4 (eBook)

Springer Cham Heidelberg New York Dordrecht London

Printed on acid-free paper

Springer is part of Springer Science+Business Media (www.springer.com)

# Preface

Vision or the ability to see is an important inborn capability of human, animals, and insects that enables them to interact with their environment. Recent advancement in computer technology led to the development of computer or machine-vision that tries to mimic vision in human and many living creatures such as insects. Attempts have been made to incorporate machine-vision in robots, unmanned aerial vehicles (UAVs), and machines with artificial intelligence. In this monograph, we consider "vision" in a more generalized sense. Here, the "eyes" that constitute a basic element of vision may correspond to sensors utilizing acoustic, electromagnetic, or other form of interaction with the environment or man-made machines. Hereafter, such "eyes" will be referred simply as "observers." In most cases, only line-ofsight observations will be considered. Given an object under observation, what an observer sees in reference to the object can be characterized by the *visible set*. In machine-vision, emphasis is placed on the interpretation of the images captured by the observer. Here, we study the characterization and geometric properties of the visible set associated with an observer and the objects under observation. Emphasis is placed on determining the extent of visual coverage of the objects under observation. To differentiate the problems under consideration here from the so-called *vision-based problems* in machine-vision, we refer to them as *visibility-based problems*.

In physical situations, the observers may be stationary or mobile. The latter case leads to path planning problems involving the selection of observation paths. The path planning problems usually arise in mobile robots and vehicles for performing certain tasks without collision with obstacles. In sensor networks, one may use sensor-equipped mobile robots that move along appropriate paths for monitoring the environment efficiently. Thus, visibility and path planning are integrated into the problem formulations. When the observation is performed over a specified or the shortest possible time-interval, it may be necessary to consider the dynamics of the observer due to inertial and frictional effects in the optimal observation-path selection. This leads to visibility-based optimal observer motion planning problems.

This monograph deals with the formulation of various visibility-based path and motion planning problems motivated by applications in appropriate mathematical framework. The solution of these problems calls for concepts and methods from many areas of applied mathematics such as computational geometry, set-covering, nonsmooth optimization, combinatorial optimization, and optimal control. Emphasis is placed on the formulation of new problems and methods of approach to these problems. Since geometry and visualization play important roles in the understanding of these problems, intuitive interpretations of the basic concepts are presented before detailed mathematical development. To fix ideas, a particular topic begins with simple cases illustrated by specific examples, and then progresses forward to more complex cases. No attempt is made to provide a survey of the enormous amount of related works pertaining to computational algorithms for computer vision, and nonvisibility-based mobile-robot path and motion planning. This monograph is directed primarily to students and researchers in engineering, computer science, and applied mathematics. An understanding of the mathematical development of the main results requires only basic knowledge of mathematical analysis, control, and optimization theories. Some exercises with various degrees of difficulty are provided at the end of the main chapters. The material presented here may serve as a portion of an introductory course or seminar on visibility-based optimal path and motion planning problems. Hopefully, this monograph is able to stimulate interest and further studies in this relatively new area.

I wish to dedicate this book to the memory of my parents who set me free at age 14 to choose my independent life's path in the United States; and to Dean Foster Strong of Caltech, whose exceptional assistance made my path through Caltech possible. Finally, I wish to acknowledge the support of the National Science Foundation and the Jet Propulsion Laboratory, Pasadena, California for the initial phase of this work.

Santa Monica, California, April 2014                    Paul Keng-Chieh Wang

# Contents

# Chapter 1
# Introduction

Visibility-based optimal path and motion planning problems arise in many real-world situations. We begin with a few examples which will be used later to illustrate the application of the theoretical results and computational algorithms to be developed in the subsequent chapters.

## 1.1 Mapping of Planetary Surface [1, 2]

In planetary exploration, one or more spacecraft or artificial satellites with onboard cameras, sensors and radar altimeters moving in the vicinity of a planet may be used to map out the planet surface and its physical properties. It is desirable to choose appropriate trajectories for the spacecraft or artificial satellite such that a specified part of the planet surface can be mapped out completely. Here, the object under observation is a 2-dimensional surface embedded in a 3-dimensional world space. The observers correspond to moving cameras and sensors with finite viewing apertures. In the case of multiple spacecraft or artificial satellites, the observation may be made in a cooperative manner so that complete surface mapping can be accomplished by using a minimal amount of non-redundant observation data. One may develop cooperative strategies based on the chosen spacecraft trajectories, or in conjunction with the motion planning task.

## 1.2 Observation of 3D-Objects [3–5]

The placement of fixed cameras for observing a 3D-object in the world space for analysis and action is a basic task in surveillance and monitoring systems. The cameras generally have finite viewing apertures, and they are mounted on fixed observation platforms. For complete visual coverage of the object, more than one camera are needed. A basic problem is to determine the minimum number of cameras and their locations for complete visual coverage of the object under observation.

© Springer International Publishing Switzerland 2015
P.K.-C. Wang, *Visibility-based Optimal Path and Motion Planning*,
Studies in Computational Intelligence 568, DOI 10.1007/978-3-319-09779-4_1

An alternate approach is to use a single moving camera to scan the surface of the 3D-object.

## 1.3 Radio Repeater Allocation [6, 7]

Modern cellular telephone and wireless communication networks make use of multiple radio or optical repeaters to cover a given service area. These repeaters receive radio or electromagnetic-wave signals from the users via line-of-sight transmission, and relay the signals to other users in the network. In the planning and design of the repeater network, it is desirable to use a minimum number of stationary repeaters to achieve complete coverage of a given service area. A basic problem is to determine the minimum number of repeaters and their locations in a specified spatial domain such that complete coverage of the service area is attained. The service area and the allowable area for repeater installation are generally not identical.

## 1.4 Imaging of Living Cells [8–10]

The identification of cancer or abnormal cells by means of computer-aided analysis of microscopic observation of a sample collection of living cells is of great interest in biomedical applications. To keep the cells alive during the observation period, they are usually immersed in a liquid medium. To obtain 3D images of the cells, more than one cameras placed on a platform outside or immersed inside the liquid medium are required. Thus, a basic problem is to determine the minimum number of cameras and their locations for a given observation platform. Recently, studies involving the interaction of living cells call for the manipulation of living cells using microscopic images. The image information may be used for the feedback control of cell movements. In this application, it is necessary to ensure that the cell properties such as geometric shapes are unaffected by the observation and actuation processes. For example, when active electromagnetic sensors such as laser-based sensors and manipulators are used for observation and actuation, the electromagnetic pressure exerted on the cell-surface produced by the sensors and actuators may affect the cell shape and structural properties.

## 1.5 Health-Monitoring and Control of Micro-distributed Systems [11]

In the health monitoring and control of micro-distributed systems such as micro-opto-electromechanical systems composed of micro-machined solid structures, it is required to observe the structural surface by means of a finite number of discrete optical sensors. An optimum design problem is to determine the minimum number of these sensors and their locations to observe the entire structural surface. This problem is akin to the well-known "Art Gallery Problem" first posed by Klee [12], i.e. determine the minimum number and locations of point guards inside an n-wall polygonal art gallery room such that every wall can be seen by at least one-guard. In the Art Gallery Problem, the observation points (locations of the guards) are in the interior or on the boundary of a polygonal spatial domain. Here, the object under observation is a surface or a 2-dimensional manifold in the 3-dimensional Euclidean space, and the observation points are restricted to another surface which does not intersect the observed one.

## 1.6 Surveillance and Exploration via Mobile Observers [13, 14]

In the surveillance of a specified terrestrial domain and exploration of a planetary surface, single or multiple Unmanned Aerial Vehicles (UAV's) and robotic rovers equipped with cameras may be used. It is desirable to find their motions such that complete visual coverage of the terrestrial domain or maximum amount of sensor data can be obtained along their corresponding paths in the spatial domain. These paths may be determined before launching the UAV's or robotic rovers based on known terrestrial data. The mobile-observer motions may also be determined in real-time based on the observed terrestrial and/or sensor data accumulated along the past path up to the present time.

## 1.7 Optical Imaging of Global Air Circulation [15, 16]

Although complex mathematical/computational models for the global air circulation in the Earth's atmosphere have been developed for weather prediction, the input data for these models are often derived from real-time optical imaging of the Earth's atmosphere exhibited by cloud formation patterns. This task can be accomplished by implementing an observer system composed of cameras attached to a set of geosynchronous and low-orbit satellites. Since the cameras have finite viewing apertures, a basic problem is to determine the minimum number of cameras and their attitudes for complete coverage of Earth's atmosphere. The satellites should be capable of communicating with each other to form a real-time global-circulation monitoring network.

## 1.8 Asteroid Observation [17, 18]

An asteroid is usually a small irregularly-shaped solid body with nearly uniform mass density. When such an asteroid enters the spatial region where the Earth's gravitational force becomes significant, it is of interest to predict its motion or path

**Fig. 1.1** Spacecraft path in world space for observing an asteroid

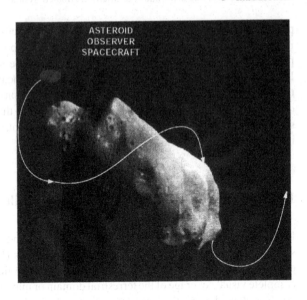

in the world or observation space, and to determine its surface and material properties. The second task can be accomplished by using one or more spacecraft equipped with cameras, radars and sensors moving in the vicinity of the asteroid. An important problem is to select the spacecraft motion so that the asteroid's surface can be mapped out completely. In the case where the observation time duration is specified, the main task is to map out the asteroid's surface as much as possible within the given observation time duration. For a single spacecraft, it is required to select a path/motion satisfying certain constraints such that maximum or complete visual coverage of the asteroid surface is attained (see Fig. 1.1). Here, the object under observation is a moving solid body in a three-dimensional world space. In this problem, a basic difficulty is that complete information about the asteroid surface is usually unavailable before launching of the spacecraft. Thus, one cannot preplan its observation path. The asteroid surface information must be acquired progressively as the spacecraft moves in the vicinity of the asteroid. We shall discuss practical methods for overcoming this difficulty later.

## 1.9 Path Planning on Structured Network

By a structured network defined on a given terrain, we mean a set of specified fixed nodes interconnected by a set of well-defined bidirectional or unidirectional paths on the terrain. Given a pair of starting and terminal nodes, it is required to find a path connecting these nodes such that an observer (e.g. a mobile robot or rover equipped with camera) attains maximum visual coverage of the terrain. This is an optimal path planning problem. For example, a tourist bus guide wishes to find the shortest sightseeing route in a city such that maximum visual coverage of the attractions can be attained. When the observation time interval is specified, the tourist bus guide may wish to plan the bus motion such that similar objective is achieved over the given time interval. This is a visibility-based optimal motion planning problem.

# References

1. D.E. Wilhelms, Geologic Mapping, Planetary Mapping, ed. by R. Greeley, R. Batson, (Cambridge Univ. Press, Cambridge, 1990), pp. 209–260
2. G. Bellucci, V. Formisano, Regional mapping of planetary surfaces with imaging spectroscopy. Planet. Space Sci. **45**, 1371–1377 (1997)
3. O. Faugeras, *Three-dimensional Computer Vision* (M.I.T. Press, Cambridge, 1992)
4. E. Trucco, A. Verri, *Introductory Techniques for 3-D Computer Vision* (Prentice Hall, Englewood Cliffs, 1998)
5. D.A. Forsyth, J. Ponce, *Computer Vision: A Modern Approach* (Prentice Hall, Englewood Cliffs, 2002)
6. R. Hoppe, P. Wertz, G. Woelfle, F. Landstorfer, Advanced ray-optical wave propagation modelling for urban and indoor scenarios including wideband properties. Eur. Trans. Telecom. **14**, 61–69 (2003)
7. P.K.C. Wang, A class of optimization problems involving set measures. Nonlinear Anal. Theor. Methods Appl. **47**, 25–36 (2001)
8. W. Choi et al., Tomographic phase microscopy. Nat. Methods **4**, 717–719 (2007)
9. P.K.C. Wang, Micro-Telerobot for manipulation of microscopic objects immersed in a liquid layer. J. Rob. Syst. **10–3**, 299–319 (1993)
10. D.J. Stephens, V.J. Allan, Light microscopy techniques for live cell imaging, science **300** (5616), 82–86 (2003)
11. P.K.C. Wang, Modelling and control of nonlinear micro-distributed systems. Nonlinear Anal. Theor. Methods Appl. **30**, 3215–3226 (1997)
12. J. O'Rourke, *Art Gallery Theorems and Algorithms* (Oxford University Press, Oxford, 1987)
13. P.K.C. Wang, Optimal motion planning for mobile observers based on maximum visibility. Dyn. Continuous, Discrete Impulsive Syst. Ser. B: Appl. Algorithms **11**, 313–338 (2004)
14. J.C. Cardema, P.K.C. Wang, Optimal Path Planning of Mobile Robots for Sample Collection on a Planetary Surface, Mobile Robots, Perception and Navigation, ed. by S. Kolski (Advanced Robotic Systems International and pro literatur Verlag, Mannendendorf, Germany, 2007), pp. 605–636
15. J.P. Peixoto, A.H. Oort, *Physics of Climate* (American Institute of Physics, New York, 1992)
16. E. Palmen, C.W. Newton, *Atmospheric Circulation Systems* (Academic Press, New York, 1969)
17. P. Ferri, Mission operations for the new Rosetta. Acta Astronaut. **58**(2), 105–111 (2006)
18. F. Gabern, W.S. Koon, J.E. Marsden, D.J. Scheeres, Binary asteroid observation orbits from a global dynamical perspective. SIAM J. Appl. Dyn. Syst. **5**(2), 252–279 (2006)

sightseeing route in a city such that maximum visual coverage of the attractions can be attained. When the observation time interval is specified, the tourist bus guide may wish to plan the bus motion such that similar objective is achieved over the given time interval. This is a visibility-based optimal motion planning problem.

## References


1. D.E. Wilhelms, Geologic Mapping, Planetary Mapping, ed. by R. Greeley, R. Batson (Cambridge Univ. Press, Cambridge, 1990), pp. 208–260
2. C. Bellucci, V. Formisano, Regional mapping of planetary surfaces with imaging spectroscopy. Planet. Space Sci. 45, 1371–1377 (1997)
3. D. [illegible]
4. R. [illegible] 1990)
5. [illegible] (1987)
6. B. Huppert, [illegible] 14 [illegible] (2003)
7. P.K.C. Wang, A class of optimization problems involving visibility [illegible] Methods Appl. 47 [illegible] (2001)
8. W. Choi et al., [illegible] (2012)
9. P.K.C. Wang, [illegible] J. Rob. Syst. 20(3) [illegible]
10. D. [illegible] (2003)
11. P.K.C. Wang, [illegible] Nonlinear Anal. Theory Methods Appl. 30 [illegible]
12. J. O'Rourke, [illegible] (Oxford, 1987)
13. P.K.C. Wang, [illegible] Dyn. Continuous Discrete Impulsive Syst. [illegible] (2004)
14. J.C. [illegible] Mobile Robots [illegible] Robotics [illegible] (2007), pp. [illegible]
15. L.S. Pontryagin, [illegible] (1962)
16. R. Bellman, [illegible] (Academic Press, New York, 19[illegible])
17. P. [illegible] SIAM [illegible] (2008)
18. [illegible] SIAM J. Appl. Dyn. Syst. 5(2), 232–279 (2006)

# Chapter 2
# Mathematical Preliminaries

We begin by formalizing various notions of visibility and what is visible to an observer in mathematical terms. These notions serve as a basis for the subsequent development.

## 2.1 Objects Under Observation

Let the *world space* $W$ be a specified subset of the $n$-dimensional Euclidean space $\mathbb{R}^n$, $n \in \{2, 3\}$ in which a vector $x$ is an ordered n-tuple $(x_1, \ldots, x_n)$ of real numbers $x_i, i = 1, \ldots, n$. The scaler product between two vectors $x$ and $x'$ in $\mathbb{R}^n$ is denoted by $\langle x, x' \rangle$. The representation of a vector $x$ with respect to a given bases $B = \{e_1, \ldots, e_n\}$ for $\mathbb{R}^n$ is denoted by the column vector $[x]$. When ambiguity does not occur, the bracket notation $[\cdot]$ is dropped for brevity. We assume that an *object under observation* $O$ is a compact (closed and bounded) connected subset of $W$ with boundary $\partial O$. The observation of the object $O$ is made from *point-observers* or *observation points* on an *observation platform* $P$ corresponding to a given subset of $W$. In what follows, we shall focus our attention mainly on the practically important case where $P \subset \overline{O^c}$ (the closure of the complement of $O$ relative to $W$). We assume that $P$ is *transparent* in the sense that it can be penetrated by the line segment or an arc connecting an observation point in $P$ and an observed point in $O$. For *opaque* or non-transparent objects, only the boundary $\partial O$ can be observed by point-observers in $P$. The difficulty associated with the visibility problems to be considered generally depends on the dimensions and geometric properties of $O$ and $P$. A few cases of special interest are given in the sequel.

**Case (i)** $W = \mathbb{R}^2$

(a) $O$ is a plane curve which can be represented by $G_f = \{(x, f(x)) \in \mathbb{R}^2 : x \in \Omega\}$, the graph of a real-valued $C_m$-function $f = f(x)$ defined on a compact interval $\Omega \subset \mathbb{R}$ (i.e. $f(\cdot) \in C_m(\mathbb{R}; \Omega)$, $m \geq 1$, the space of all real-valued functions having continuous derivatives on $\Omega$ up to the $m$th order). The observations of the

© Springer International Publishing Switzerland 2015
P.K.-C. Wang, *Visibility-based Optimal Path and Motion Planning*,
Studies in Computational Intelligence 568, DOI 10.1007/978-3-319-09779-4_2

**(a)**                                                              **(b)**

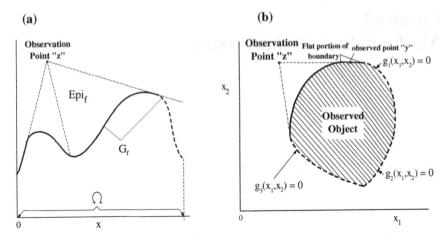

**Fig. 2.1** Cases (i)(a) and (i)(b). The solid and dashed curves correspond to points of the object that are visible and invisible from the observation point $z$ respectively

curve are made from points $z = (x, w) \in \mathcal{P} \subset \text{Epi}_f \overset{\text{def}}{=} \{(x', w') \in \Omega \times \mathbb{R} : w' \geq f(x')\}$, the *epigraph* of $f$ (See Fig. 2.1a). We may also consider observations from points below the curve $G_f$, i.e. $z \in \{(x', w') \in \Omega \times \mathbb{R} : w' < f(x')\}$.

(b) $\mathcal{O}$ is an opaque object represented by a compact connected subset of $\mathcal{W}$. It can be described by $\bigcap_{k \in \{1,\dots,K\}} \{x \in \mathcal{W} : g_k(x) \leq 0\}$, and has interior points and a piecewise smooth boundary $\partial\mathcal{O}$, where the $g_k$'s are specified functions in $C_m(\mathbb{R}; \mathcal{W})$, $m \geq 1$. The point-observers or observation points lie in $\mathcal{O}^c$ (See Fig. 2.1b).

**Case (ii)** $\mathcal{W} = \mathbb{R}^3$

(a) $\mathcal{O}$ is an opaque smooth surface described by $G_f$, the graph of a real-valued $C_m$-function $(m \geq 1) f = f(x), x = (x_1, x_2) \in \Omega$, a compact connected subset of $\mathbb{R}^2$ (See Fig. 2.2). The observation points $z = (x, w)$ are located in $\text{Epi}_f \subset \mathbb{R}^3$.

(b) $\mathcal{O}$ is an opaque solid body represented by a nonempty compact connected subset of $\mathbb{R}^3$ with interior points and a piecewise smooth boundary $\partial\mathcal{O}$ which may be expressed as a level set of a real-valued continuous function $g = g(x)$ defined on $\mathbb{R}^3$, i.e. $\mathcal{O} = \{x \in \mathbb{R}^3 : g(x) = \alpha\}$ or $g^{-1}(\alpha)$, where $\alpha \in \mathbb{R}$. As in Case (i)(b), the object $\mathcal{O}$ may also be described by $\bigcap_{k \in \{1,\dots,K\}} \{x \in \mathcal{W} : g_k(x) \leq 0\}$, where the $g_k$'s are specified functions in $C_m(\mathbb{R}; \mathcal{W})$. The observation points lie in $\mathcal{O}^c$.

In physical situations, the object under observation in Case (ii)(a) may correspond to the spatial profile of a terrain (e.g. Mars surface) under observation by a robot camera. Thus the region below the surface (i.e. the complement of $\text{Epi}_f$ relative to $\Omega \times \mathbb{R}$) is irrelevant. The object under observation in Case (ii)(b) may correspond to an asteroid as described in Chap. 1, a nano-particle, or a product under inspection in

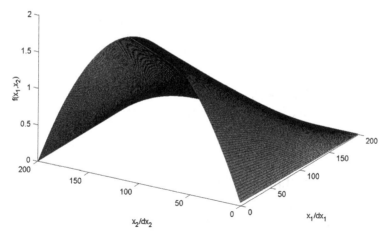

**Fig. 2.2**  The object $\mathcal{O}$ under observation is an opaque surface in $\mathbb{R}^3$ described by $G_f$. The observation points are located above the surface or in $\text{Epi}_f$.

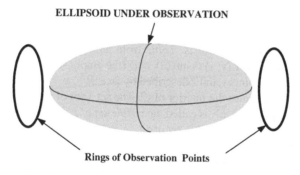

**Fig. 2.3**  $\partial\mathcal{O}$ is an ellipsoid, and $\mathcal{P}$ consists of two rings

a manufacturing plant. Figure 2.3 shows an example where $\partial\mathcal{O}$ is an ellipsoid, and the observation platform $\mathcal{P}$ is composed of two rings.

**Case (iii)** $\mathcal{W} = \mathbb{R}^n$, $n = 2, 3$

Cases (i) and (ii) do not involve the time of observation. Moreover, the shapes of the objects $\mathcal{O}$ under observation are assumed to be time-invariant over the observation period. This assumption may be invalid in many physical situations. For example, if $\mathcal{O}$ is a closed bounded subset of Earth's solid surface, its shape may vary significantly over the observation period during an Earth quake. Another example is the heating or cooling of a solid body $\mathcal{O} \subset \mathbb{R}^3$ whose shape may change significantly over the observation period. In these cases, we may characterize the objects under observation by an *one-parameter (time t) family of objects* $\mathcal{O}_t$, i.e. $\mathcal{F}_{\mathcal{O}} = \{\mathcal{O}_t : t \in I_T\}$, where $I_T$ is a specified observation time-interval.

*Example 2.1*  $\mathcal{F}_O$ is an one-parameter family of 2-dimensional surfaces in $\mathbb{R}^3$ described by $\mathcal{O}_t = \{(x_1, x_2, f(t, x_1, x_2)) \in \mathbb{R}^3 : sf(t, x_1, x_2) = \sin(2\pi t)\sin(\pi x_1) \sin(2\pi x_2); (x_1, x_2) \in \Omega\}$, where $t \in I_T = [0, 10]$ and $\Omega = \{(x_1, x_2) \in \mathbb{R}^2 : x_1^2 + x_2^2 \le 1\}$.

*Example 2.2*  $\mathcal{F}_O$ is an one-parameter family of pulsating spherical balls in $\mathbb{R}^3$ described by $\mathcal{O}_t = \{(x_1, x_2, x_3) \in \mathbb{R}^3 : x_1^2 + x_2^2 + x_3^2 \le (1 + 0.25\sin(t))^2\}$, $t \in I_T = [0, 2\pi]$.

## 2.2 Notions of Visibility [1, 2]

First, we consider objects $\mathcal{O}$ with time-invariant shapes as described in Cases (i) and (ii). We introduce the notion of visibility of a point $y$ on $\partial\mathcal{O}$ from an observation point $z$, assuming no other objects or obstacles are in the world space $\mathcal{W}$. Unless stated otherwise, the object under observation is assumed to be opaque, and the observation point $z \in \overline{\mathcal{O}^c}$. The visibility of the point under observation is based on line-of-sight from the observation point $z$.

**Definition 2.1**  A point $y \in \partial\mathcal{O}$ is *visible* from an observation point $z \in \overline{\mathcal{O}^c}$, if (i) $\mathsf{L}(y, z) \subset \mathcal{W}$, and (ii) $\mathsf{L}(y, z) \cap \text{int}(\mathcal{O}) = \phi$ (the empty set), where $\mathsf{L}(y, z)$ is the line segment joining points $y$ and $z$ described by $\{x \in \mathbb{R}^3 : x = \lambda y + (1 - \lambda)z, 0 < \lambda < 1\}$, and $\text{int}(\mathcal{O})$ denotes the interior of $\mathcal{O}$. The set $\mathcal{V}(z)$ of all points $y \in \partial\mathcal{O}$ that are visible from a point $z \in \overline{\mathcal{O}^c}$ is called the *visible set of $z$*. The complement of $\mathcal{V}(z)$ relative to $\partial\mathcal{O}$ is called the *invisible set of $z$*.

Condition (ii) implies that $\mathsf{L}(y, z)$ does not penetrate into the interior of $\mathcal{O}$, and $\mathsf{L}(y, z) \cap \partial\mathcal{O}$ may be nonempty. In the case where diffraction phenomenon occurs, we may replace $\mathsf{L}(y, z)$ in Definition 2.1 by a smooth arc (in a given class) connecting the points $y$ and $z$. Figure 2.1 shows the visible and invisible sets of a point $z$ for observing a plane curve and a compact connected set in $\mathbb{R}^2$ indicated by solid and dashed curves respectively. Note that the boundary of the object in Fig. 2.1b has a flat portion. Thus, for the indicated observation point $z$ and observed point $y$, $\mathsf{L}(y, z) \cap \partial\mathcal{O}$ is nonempty. The visible and invisible sets of an observation point for an opaque solid object in $\mathbb{R}^3$ are illustrated by Fig. 2.4.

*Remark 2.1*  When a point-observer $z$ has finite viewing aperture such as a camera, we may define the visible set of the point-observer $z$ as

$$\mathcal{V}(z) = \{y \in (\mathcal{C}_z \cap \partial O) : \lambda z + (1 - \lambda)y \notin O \text{ for all } \lambda \in \,]0, 1[\}, \qquad (2.1)$$

where $\mathcal{C}_z$ is the *cone of visibility* associated with the point-observer $z$ at its vertex, and has a finite viewing-aperture angle as illustrated in Fig. 2.5. In the case of a camera, its visible set may be enlarged by rotating the camera about the vertex of its viewing-aperture cone. This approach may be useful when the object under observation does not change its shape significantly during the rotation period. Throughout this book,

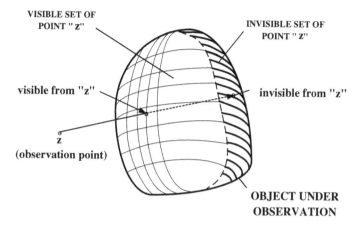

VISIBLE SET OF
POINT "z"

INVISIBLE SET OF
POINT "z"

visible from "z"

invisible from "z"

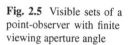

z

(observation point)

OBJECT UNDER
OBSERVATION

**Fig. 2.4** Visible and invisible sets of an observation point for an opaque solid object in $\mathbb{R}^3$

**Fig. 2.5** Visible sets of a
point-observer with finite
viewing aperture angle

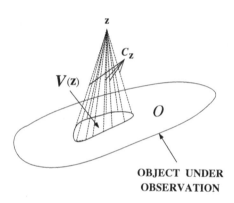

z

$C_{\mathbf{z}}$

$V(\mathbf{z})$

$O$

OBJECT UNDER
OBSERVATION

unless stated otherwise, no restriction is imposed on the viewing aperture of a point-observer.

*Remark 2.2* In Definition 2.1, visibility is defined in terms of line-of-sight observations. In practical situations, observations may be accomplished using more complex devices. For example, the observation of an object $\mathcal{O}$ may be made using a reflective surface $R_f$ as illustrated in Fig. 2.6. Here, a point $y \in \partial\mathcal{O}$ is visible from a point-observer $z \in \overline{\mathcal{O}^c}$, if there exists a ray composed of incident and reflected rays connecting $z$ and $y$. The direction of the reflected ray is determined by the usual Law of Reflection in optics (i.e. angle of incidence equals the angle of reflection). In this work, we only consider line-of-sight observations.

Figure 2.7 shows the visible sets from two different observation points for Case (ii)(a) in which the observed object is a surface in $\mathbb{R}^3$. We may associate this case with a mountaineer (the observer) climbing up a mountain surface. When the mountaineer is at position "a", his visible set is limited to only one side of the mountain surface. But when the climber reaches position "b", his visible set suddenly expands to both

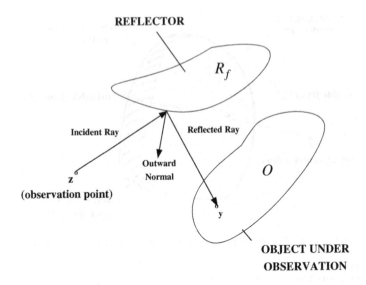

**Fig. 2.6**  Observation of object $\mathcal{O}$ from a point-observer via a reflective surface $R_f$

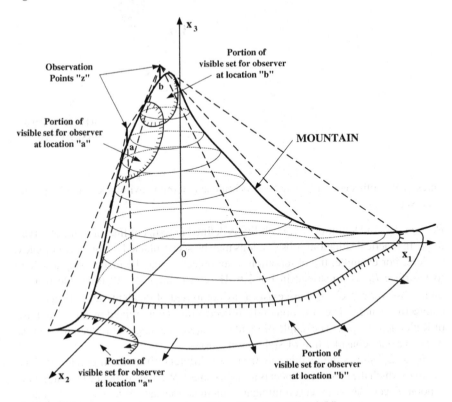

**Fig. 2.7**  Visible sets of an observer at two different locations "a" and "b"

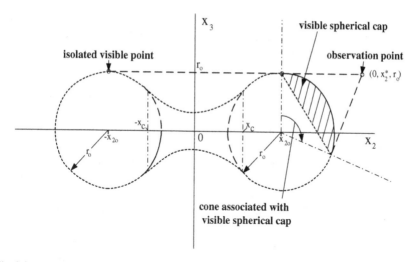

**Fig. 2.8** Projection of $\mathcal{V}((0, x_2^*, r_o))$ onto the $(x_2, x_3)$-plane

sides of the mountain surface as if the world is below his feet (an exhilarating moment experienced by many mountaineers!).

In general, $\mathcal{V}(z)$ may be composed of disconnected sets, isolated points and arcs as illustrated by the following example:

*Example 2.3* Let $B = \{e_1 = (1, 0, 0), e_2 = (0, 1, 0), e_3 = (0, 0, 1)\}$ be the orthonormal bases for $\mathbb{R}^3$. A point $x \in \mathbb{R}^3$ can be specified by the ordered triplet $(x_1, x_2, x_3)$ or its representation with respect to $B$ given by the column vector $[x_1, x_2, x_3]^T$. Consider a solid body $\mathcal{O}$ in $\mathbb{R}^3$ whose boundary surface $\partial\mathcal{O}$ is formed by revolving the plane curve $C$ composed of three circular arcs about the $x_2$-axis (see Fig. 2.8), where $C = \{(x_1, x_2, x_3) \in \mathbb{R}^3 : x_1 = 0; x_3 = \sqrt{r_o^2 - (x_2 - x_{2o})^2}$ if $|x_2| > x_c$; $x_3 = \sqrt{r_c^2 - x_2^2} + x_{3o}$ if $|x_2| \leq x_c\}$, with $r_o$ and $r_c = \sqrt{x_{2o}^2 + x_{3o}^2} - r_o > 0$ being the radii of curvature of the circular arcs, and $x_c = x_{2o} r_c / \sqrt{x_{2o}^2 + x_{3o}^2}$. For the observation point $z = (x_1, x_2, x_3) = (0, x_2^*, r_o)$ with $x_2^* > x_{2o} > 0$, its visible set is given by $\mathcal{V}(z) = \hat{C} \cup \{(0, -x_{2o}, r_o)\}$, where $\hat{C}$ is a spherical cap associated with the sphere centered at $x = (0, x_{2o}, 0)$ with radius $r_o$, described by

$$\hat{C} = \{(x_1, x_2, x_3) \in \mathbb{R}^3 : x_1^2 + (x_2 - x_{2o})^2 + x_3^2$$
$$= r_o^2; (x_2 - x_{2o}) + r_o x_3 / (x_2^* - x_{2o}) \geq r_o^2 / (x_2^* - x_{2o})\}. \qquad (2.2)$$

When more than one object or obstacles are present in the world space $\mathcal{W}$, there may exist observation points from which no points on $\partial\mathcal{O}$ are visible.

*Remark 2.3* In many practical situations, attention may be focused on certain special subsets of the visible sets of the observation points rather than the entire visible sets. These subsets may be characterized by special properties. For example, let $\mathcal{V}(z)$ be the visible set of a point $z \in \mathcal{P}$ for observing a 2D-surface $\mathcal{O} = \{(x, f(x)) \in \mathbb{R}^3 : x \in \Omega \subset \mathbb{R}^2\}$, where $f$ is a real-valued continuous function of $x \in \Omega$. The subsets $\tilde{\mathcal{V}}(z)$ of $\mathcal{V}(z)$ of special interest correspond to certain specified level-sets of $\mathcal{V}(z)$, i.e. $\tilde{\mathcal{V}}(z) = \{(x, f(x)) \in \mathcal{V}(z) : f(x) = c, x \in \Omega\}$, where $c$ is a specified real number. Thus, we may describe this situation by defining a set-valued mapping $\Upsilon$ on $\mathcal{P}$ into $2^{\mathcal{V}(z)}$ such that $z \to \tilde{\mathcal{V}}(z)$.

Let $n(x)$ denote the outward unit normal at a point $x \in \partial\mathcal{O}$, and $T_x(\partial\mathcal{O})$ the corresponding tangent plane at $x$. We assume that the boundary of $\mathcal{O}$ is describable by the level set of a smooth function $f = f(x)$ defined on a compact set $\Omega \subset \mathcal{W}$, i.e. $\partial\mathcal{O} = \{x \in \Omega : f(x) = \alpha\}$ for some real number $\alpha$. Then, the outward unit normal at $x$ is defined by $n(x) = \nabla f(x)/\|\nabla f(x)\|$, where $\nabla f(x)$ denotes the gradient of $f$ at $x \in \partial\mathcal{O}$. A special observation platform is the dilated $\partial\mathcal{O}$ such that at any point $x \in \partial\mathcal{O}$, the corresponding observation point $z \in \mathcal{P}$ is at a given height $h(x) > 0$ along $n(x)$ above $x$, where $h = h(x)$ is a specified positive continuous function of $x$ defined on $\partial\mathcal{O}$. The corresponding observation platform (denoted hereafter by $\mathcal{P}_h$) is described by $\{z \in \mathcal{W} : z = x + h(x)n(x), x \in \partial\mathcal{O}\}$, a dilation of $\partial\mathcal{O}$ in the directions of the surface outward normals with increment $h(\cdot)$. When $h$ is a positive constant, we have the practically important case of a *constant-height observation platform*. Note that since $\partial\mathcal{O}$ is a compact set in $\mathcal{W}$, so is the observation platform $\mathcal{P}_h$.

Given an object $\mathcal{O}$ in $\mathcal{W}$, and a set $\mathcal{P}$ of observation points $z$, the visible set of $\mathcal{P}$ (denoted by $\mathcal{V}(\mathcal{P})$) is given by $\bigcup_{z \in \mathcal{P}} \mathcal{V}(z)$.

**Definition 2.2** The object $\mathcal{O}$ is said to be *partially visible from* $\mathcal{P}$, if $\mathcal{V}(\mathcal{P})$ is a nonempty proper subset of $\partial\mathcal{O}$. If $\mathcal{V}(\mathcal{P}) = \partial\mathcal{O}$, then $\mathcal{O}$ is said to be *totally visible from* $\mathcal{P}$.

First, consider the case where $\mathcal{P} = \mathcal{O}^c$. A basic question pertaining to total visibility can be posed as follows: Given an object $\mathcal{O} \subset \mathcal{W}$, does there exist an observation point $z \in \mathcal{O}^c$ at a finite distance from $\mathcal{O}$ such that $\mathcal{O}$ is totally visible? If not, is totally visibility of $\mathcal{O}$ attainable by a finite set of point-observers $z^{(i)} \in \mathcal{O}^c$, $i = 1, \ldots, N$, each located at a finite distance from $\mathcal{O}$? Here, the distance between $z^{(i)}$ and $\mathcal{O}$ is defined by

$$\rho(z^{(i)}, \mathcal{O}) \overset{\text{def}}{=} \inf\{\|z^{(i)} - x\| : x \in \mathcal{O}\}, \quad i = 1, \ldots, N, \tag{2.3}$$

where $\| \cdot \|$ denotes the Euclidean norm.

The following proposition provides the answer to the foregoing question for observed objects $\mathcal{O}$ corresponding to Cases (i)(a) and (ii)(a):

**Proposition 2.1** *For an object $\mathcal{O}$ corresponding to a plane curve or a smooth surface described in Cases (i)(a) or (ii)(a) in Sect. 2.1 respectively, there exists an observation point $z \in \text{Epi}_f$ from which $\mathcal{O}$ is totally visible.*

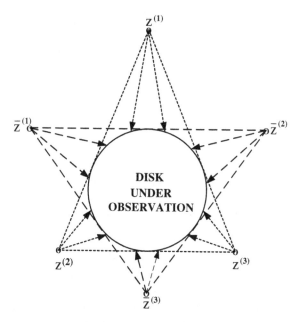

**Fig. 2.9** Total visibility of a circular disk is attainable by a set of three distinct observation points located at the vertices of any triangle that inscribes the disk, e.g. $\{z^{(1)}, z^{(2)}, z^{(3)}\}$ and $\{\bar{z}^{(1)}, \bar{z}^{(2)}, \bar{z}^{(3)}\}$

A proof for the above result is included in the proof of Lemma 3.1 in Chap. 3. This result is consistent with intuition that a smooth surface or terrain without folds is totally visible if the observation point is sufficiently high above the surface.

Next, consider Case (i)(b) in Sect. 2.1 where the object $\mathcal{O}$ is described by a compact connected subset of $\mathcal{W} = \mathbb{R}^2$ with interior points and a piecewise smooth boundary $\partial \mathcal{O}$. To fix ideas, let the object $\mathcal{O}$ under observation be a circular disk. Evidently, total visibility of the disk or its circular boundary $\partial \mathcal{O}$ can be attained by three point-observers located at the vertices of any triangle that inscribes the disk (see Fig. 2.9). Moreover, the visible set of a point-observer at a finite distance from the disk is always a proper subset of a boundary semi-circle. Therefore total visibility of a circular disk cannot be attained using less than three point-observers that are at finite distances from the disk. Now, one may ask whether at least three point-observers are necessary for total visibility of any compact simply connected object in $\mathbb{R}^2$. A little reflection leads one to conclude that for some compact connected objects in $\mathbb{R}^2$, only two point-observers are required for total visibility. For example, a $n$-sided polygon requires only two point-observers at finite distances outside the polygon for total visibility. However, its boundary is piecewise smooth.

**Proposition 2.2** *For an object $\mathcal{O}$ described by a compact connected subset of $\mathbb{R}^2$ with interior points and a piecewise smooth boundary described in Cases (i)(b) in Sect. 2.1, at least two point-observers located in $\mathcal{O}^c$ are required for total visibility of $\mathcal{O}$.*

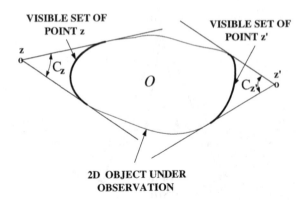

**Fig. 2.10** Illustration of at least two point observers are needed for total visibility

*Proof* Since $\mathcal{O}$ is compact, its diameter $\mathrm{diam}(\mathcal{O}) \overset{\mathrm{def}}{=} \max\{\|x - x'\| : x, x' \in \mathcal{O}\}$ is finite. For any observation point $z \in \mathcal{O}^c$, we can construct a cone $C_z \subset \mathbb{R}^2$ with its vertex at $z$ such that $\mathcal{O} \subset C_z$ and the boundary lines of $C_z$ are tangent to $\partial\mathcal{O}$ as illustrated by Fig. 2.10. Clearly, it is impossible to attain total visibility of $\mathcal{O}$ by any single observation point $z \in \mathcal{O}^c$. Thus, we can partition $\partial\mathcal{O}$ into two subsets, $\mathcal{V}(z)$ and $\partial\mathcal{O} - \mathcal{V}(z)$. The total visibility of the latter subset requires at least one other point-observer.                                                    □

**Proposition 2.3** *Let $\mathcal{O}$ be an object under observation described by a compact connected subset of $\mathbb{R}^n$, $n \in \{1, 2, 3\}$, such that $\partial\mathcal{O}$ is totally visible from a given observation platform $\mathcal{P} \subset \overline{\mathcal{O}^c} \subset \mathbb{R}^n$. Let $T$ be any nonsingular linear mapping from $\mathbb{R}^n$ onto itself. Then, $\partial(T\mathcal{O})$ is totally visible from $T\mathcal{P}$.*

*Proof* From the assumption that $\partial\mathcal{O}$ is totally visible from a given observation platform $\mathcal{P}$, for any point $y \in \partial\mathcal{O}$, there exists a point $z \in \mathcal{P}$ such that the line segment $L(z, y) \cap \mathrm{int}(\mathcal{O}) = \phi$ (the empty set). Now, for any nonsingular linear transformation $T$ on $\mathbb{R}^n$ onto itself, the image of $L(z, y)$ under the mapping $T$ is the line segment $L(Tz, Ty)$ which does not intersect $\mathrm{int}(T\mathcal{O})$. By definition, the point $Ty \in \partial(T\mathcal{O})$ is visible from $Tz \in T\mathcal{P}$. It follows that $\partial(T\mathcal{O})$ is totally visible from $T\mathcal{P}$.                                                    □

The significance of Proposition 2.3 is that once the total visibility of an object $\mathcal{O}$ from a given platform $\mathcal{P}$ is established, one can generate families of objects with different shapes via nonsingular linear transformations $T$ such that the pairs $\{T\mathcal{O}, T\mathcal{P}\}$ have the total visibility property.

The foregoing elementary observations suggest the following constructive approach to the determination of point-observers for a special class of simply connected compact objects in $\mathbb{R}^2$ that are related to a circular disk by a continuous radial deformation as illustrated in Fig. 2.11. Note that there exist compact connected objects in $\mathbb{R}^2$ that are not transformable from a circular disk via radial deformations (see Fig. 2.12 for a few examples). For this special class of objects $\mathcal{O}$, we assume that their boundary curves $\partial\mathcal{O}$ can be expressed in the form:

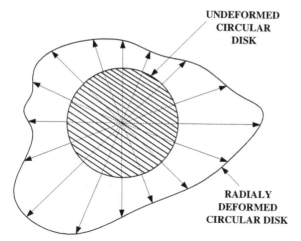

**Fig. 2.11** Sketch of a radially deformed circular disk in $\mathbb{R}^2$

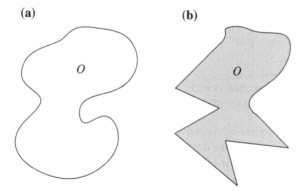

**Fig. 2.12** Examples of compact connected objects in $\mathbb{R}^2$ that are not transformable from a circular disk via radial deformations. **a** object with smooth boundary; **b** object with nonsmooth boundary

$r = r(\theta)$, where $r$ (the radial distance from the disk center) is a positive $C_m$-function ($m \geq 1$) of the angle $\theta \in [0, 2\pi[$. Thus, for each point $x \in \partial\mathcal{O}$, there exists a unique outward unit normal $n(x)$ with its corresponding tangent line described by $L(x) \stackrel{\text{def}}{=} \{x' \in \mathbb{R}^2 : \langle n(x), x' - x \rangle = 0\}$, where $\langle \cdot, \cdot \rangle$ denotes the inner product on $\mathbb{R}^2$. Suppose we can find a pair of adjacent points $a, b \in \partial\mathcal{O}$ such that their corresponding tangent lines $L(a)$ and $L(b)$ intersect at a point $z \in \mathcal{O}^c$, then the arc $\mathcal{A}(a, b) \subset \partial\mathcal{O}$ connecting $a$ and $b$ is visible from $z$ (see Fig. 2.13), provided that $L(a)$ and $L(b)$ are lines of support to $\mathcal{O}$ with the *separation property*: $\mathcal{O}$ lies in the negative half-spaces $H^-(a) \stackrel{\text{def}}{=} \{x \in \mathbb{R}^2 : \langle n(a), x - a \rangle \leq 0\}$ and $H^-(b) \stackrel{\text{def}}{=} \{x \in \mathbb{R}^2 : \langle n(b), x - b \rangle \leq 0\}$ respectively. Moreover, $L(a) \cap \mathcal{O} = \{a\}$ and $L(b) \cap \mathcal{O} = \{b\}$. Under the foregoing conditions, we can determine the point-observers for this class of objects by first

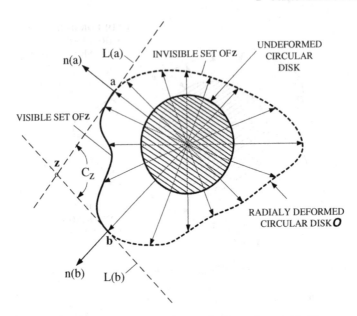

**Fig. 2.13** Construction of point-observers via intersecting lines of support to $\mathcal{O}$

determining the intersection points $z$ of the lines of support for $\mathcal{O}$ for points along $\partial\mathcal{O}$, and then select a subset of these points $z$ to attain total visibility.

*Remark 2.4* There exist objects in the foregoing special class that do not have intersecting lines of support with the desired separation property. In Fig. 2.13, the object $\mathcal{O}$ has only one pair of points on $\partial\mathcal{O}$ whose supporting lines have the desired separation property. Therefore total visibility of $\mathcal{O}$ cannot be attained by the point of intersection of these lines. If $\mathcal{O}$ is convex with a smooth boundary, then the existence of such lines of support is guaranteed by the Hahn-Banach Theorem (Separation form) [3]. Figure 2.14 shows an object in the above-mentioned class for which total visibility is attainable by three point-observers determined by the proposed approach.

Now, consider Case (ii)(b) in Sect. 2.1 where the observed object $\mathcal{O}$ is a solid body in $\mathbb{R}^3$. Following the line-of-thinking in the foregoing discussion, let $\mathcal{O}$ be a closed spherical ball $\bar{B}(0; R) = \{x = (x_1, x_2, x_3) \in \mathbb{R}^3 : \|x\| \leq R\}$ centered at the origin of $\mathbb{R}^3$ with radius $R$. For any point-observer $z \in \bar{B}(0; R)^c$ with $R < \|z\| < \infty$, its visible set $\mathcal{V}(z)$ is the spherical cap $\hat{C}(z)$ described by

$$\hat{C}(z) \stackrel{\text{def}}{=} \mathcal{S}(0; R) \cap H^+(R^2/\|z\|), \tag{2.4}$$

where $H^+(R^2/\|z\|)$ denotes the half-space $\{x \in \mathbb{R}^3 : \langle x, z/\|z\| \rangle \geq R^2\|z\|\} = \{x \in \mathbb{R}^3 : \langle x, z \rangle \geq R^2\}$ and $\langle \cdot, \cdot \rangle$ denotes the scalar product. We note that as $\|z\| \to \infty$, $\hat{C}(z)$ tends to a hemisphere of $\mathcal{S}(0; R)$.

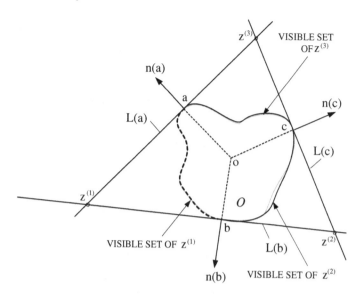

**Fig. 2.14** Example of a 2D object which is totally visible by a set of three point-observers obtained by intersecting lines of support to $\mathcal{O}$

In view of the result for total visibility of a circular disk in $\mathbb{R}^2$, we may conjecture that $\bar{B}(0; R)$ or its boundary sphere $\mathcal{S}(0; R) = \{x \in \mathbb{R}^3 : \|x\| = R\}$ is totally visible from a set of four point-observers at the vertices $z^{(i)}$ of a regular tetrahedron with edge $a$ inscribing $\mathcal{S}(0; R)$ as shown in Fig. 2.15. The vertices of the tetrahedron are given by

$$
\begin{aligned}
z^{(1)} &= (0, 0, h - R) = (0, 0, 3)R, \\
z^{(2)} &= (0, -2L/3, -R) = (0, -2\sqrt{2}, -1)R, \\
z^{(3)} &= (a/2, L/3, -R) = (\sqrt{6}, \sqrt{2}, -1)R, \\
z^{(4)} &= (-a/2, L/3, -R) = (-\sqrt{6}, \sqrt{2}, -1)R,
\end{aligned}
\tag{2.5}
$$

where $h = a\sqrt{6}/3$ is the altitude and $L = a\sqrt{3}/2$ the slant height of the regular tetrahedron. $R = h/4 = a\sqrt{6}/12$ is the radius of the inscribed sphere $S(0; R)$ and $a = 2\sqrt{6}R$ is the edge of the regular tetrahedron. To verify the validity of the foregoing conjecture, we first compute the visible sets $\mathcal{V}(z^{(i)})$ described by spherical caps:

$$
\begin{aligned}
\hat{C}(z^{(1)}) &= S(0; R) \cap \{(x_1, x_2, x_3) \in \mathbb{R}^3 : x_3 \geq R/3\}, \\
\hat{C}(z^{(2)}) &= S(0; R) \cap \{(x_1, x_2, x_3) \in \mathbb{R}^3 : -2\sqrt{2}x_2 - x_3 \geq R\}, \\
\hat{C}(z^{(3)}) &= S(0; R) \cap \{(x_1, x_2, x_3) \in \mathbb{R}^3 : \sqrt{6}x_1 + \sqrt{2}x_2 - x_3 \geq R\}, \\
\hat{C}(z^{(4)}) &= S(0; R) \cap \{(x_1, x_2, x_3) \in \mathbb{R}^3 : -\sqrt{6}x_1 + \sqrt{2}x_2 - x_3 \geq R\}.
\end{aligned}
\tag{2.6}
$$

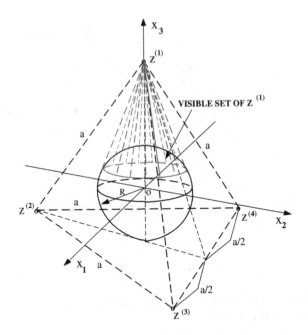

**Fig. 2.15** Four point-observers at the vertices of a regular tetrahedron inscribing a spherical ball with radius $R$

For total visibility, the visible sets of the four observation points must cover the sphere, i.e. $\bigcup_{i=1}^{4} \hat{C}(z^{(i)}) = \mathcal{S}(0; R)$. Evidently, the sphere $\mathcal{S}(0; R)$ cannot be partitioned into four *disjoint* spherical caps. Consider $\hat{C}(z^{(1)})$ whose lower boundary is the circle:

$$C_1 \overset{\text{def}}{=} \{x = (x_1, x_2, x_3) \in \mathbb{R}^3 : x_1^2 + x_2^2 = 8R^2/9, x_3 = R/3\}.$$ It can be verified by straightforward computation that

$$C_1 \cap \hat{C}(z^{(2)}) = \{(x_1, x_2, x_3) \in \mathbb{R}^3 : x_1^2 + x_2^2 = 8R^2/9; x_2 \leq -\sqrt{2}R/3; x_3 = R/3\},$$

$$C_1 \cap \hat{C}(z^{(3)}) = \{(x_1, x_2, x_3) \in \mathbb{R}^3 : x_1^2 + x_2^2 = 8R^2/9; x_1 \geq 0; x_2 \geq -\sqrt{2}R/3; x_3 = R/3\}, \quad (2.7)$$

$$C_1 \cap \hat{C}(z^{(4)}) = \{(x_1, x_2, x_3) \in \mathbb{R}^3 : x_1^2 + x_2^2 = 8R^2/9; x_1 \leq 0; x_2 \geq -\sqrt{2}R/3; x_3 = R/3\}.$$

The arcs $C_1 \cap \hat{C}(z^{(i)})$, $i = 2, 3, 4$ are subsets of the circle $C_1$ (see Fig. 2.16), and $\bigcup_{i=2}^{4}(C_1 \cap \hat{C}(z^{(i)})) = C_1$. Evidently, since the tetrahedron is regular, we can repeat the foregoing computations for $C_j \cap \hat{C}(z^{(i)})$ for $j \neq 1$ and $i \neq j$ and arrive at a similar conclusion, where $C_j$ is the circle associated with the spherical cap corresponding to the visible set of $z^{(j)}$. Thus, we have verified the truth of our conjecture that total visibility is attainable by the four observation points $z^{(i)}$, $i = 1, \ldots, 4$ at the vertices of any inscribing regular tetrahedron. At this point, we may ask whether total visibility is attainable by other 4-point-observer configurations. Since the visible set of a point-observer tends to a hemisphere of $\mathcal{S}(0; R)$ as the distance between the

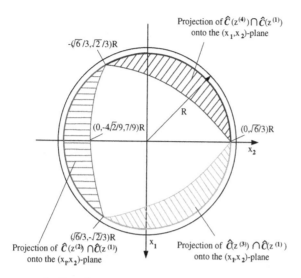

**Fig. 2.16** Projection of $\hat{C}(z^{(1)}) \cap \hat{C}(z^{(i)})$, $i = 2, 3, 4$ onto the $(x_1, x_2)$-plane

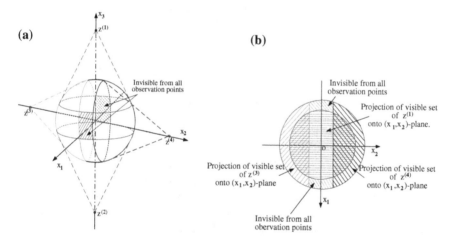

**Fig. 2.17** Visible sets of a sphere attainable by four point observers. **a** Point-observer configuration; **b** Projections of visible sets of $z^{(i)}$, $i = 1, \ldots, 4$ onto the $(x_1, x_2)$-plane

observation point $z$ and $\mathcal{S}(0; R)$ tends to infinity, it follows that the 4-point-observers at the vertices of any *dilated* inscribing regular tetrahedron (i.e. equal extension of the point-observers along the directions of the vertices) also leads to total visibility.

We may consider other 4-point-observer configurations such as the one shown in Fig. 2.17. Here, the non-overlapping portions of the visible sets of all four observation points have the same surface area. We note there remain two patches on the sphere that are invisible from any point-observer. The areas of these invisible patches tend to zero as the distances between the point-observers and the sphere tend to infinity. Evidently,

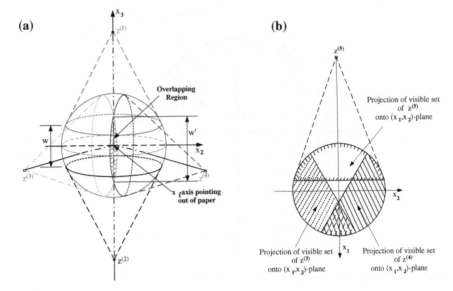

**Fig. 2.18** Total visibility of a sphere attainable by five-point observers. **a** Point-observer configuration; **b** Projections of visible sets of $z^{(i)}$, $i = 3, 4, 5$ onto the $(x_1, x_2)$-plane

total visibility of the sphere is not attainable by this nonsymmetric 4-point-observer configuration.

Consider another nonsymmetric configuration consisting of five point-observers. We shall show by construction that total visibility is attainable by a 5-point-observer configuration shown in Fig. 2.18. It consists of two point-observers $z^{(1)}$ and $z^{(2)}$ located symmetrically about the origin on the $x_3$-axis at $(0, 0, \pm d)$. Their visible sets are spherical caps separated by a ring with width $w$. This ring can be made visible by three point-observers whose visible sets overlap each other as indicated by the projection of the cutting planes associated with the spherical caps onto the $(x_1, x_2)$-plane as shown in Fig. 2.18b. To attain total visibility, the overlapping height $w$ (see Fig. 2.18a) must be greater than $w'$. This can be accomplished by setting the point-observers $z^{(3)}$ and $z^{(4)}$ sufficiently far apart.

*Remark 2.5* We note that Proposition 2.2 remains valid for objects in $\mathcal{W} = \mathbb{R}^3$ described in Case(ii)(b) in Sect. 2.1. Also, for certain objects, total visibility is attainable by two point-observers. For example, only two point-observers are required for total visibility of the object formed by a cone and a portion of a spherical ball as shown in Fig. 2.19.

In applications, a basic question is: Given $\mathcal{P}$ and $\mathcal{O}$, is $\mathcal{O}$ totally visible from $\mathcal{P}$? In general, $\mathcal{P}$ may be a finite set of observation points or the union of disjoint subsets of $\mathcal{O}^c$. In what follows, we assume that the observation points $z$ are restricted to a transparent observation platform $\mathcal{P}$ in $\mathcal{O}^c$. For $\mathcal{P} = \mathcal{P}_h$, the point $x \in \partial\mathcal{O}$ described by $x = z - h(x)n(x)$ is always visible from $z$. Hence $\mathcal{V}(z)$ is nonempty. Thus, we may

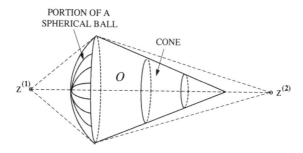

PORTION OF A
SPHERICAL BALL

CONE

$O$

$z^{(1)}$

$z^{(2)}$

**Fig. 2.19** 3-D object formed by a cone and a portion of a spherical ball. Total visibility is attained by two point-observers $z^{(1)}$ and $z^{(2)}$

regard $z \to \mathcal{V}(z)$ as a set-valued mapping on $\mathcal{P}_h$ into $2^{\partial\mathcal{O}}$, the space of all nonempty compact subsets of $\partial\mathcal{O}$. In general, this mapping is discontinuous with respect to the metric $\rho$ on $\mathcal{P}_h$, and Hausdorff metric $\rho_H$ on $2^{\partial\mathcal{O}}$ defined by

$$\rho_H(A, B) \stackrel{\text{def}}{=} \max\{\max\{\rho(x, B) : x \in A\}, \max\{\rho(x', A) : x' \in B\}\}, \tag{2.8}$$

where

$$\rho(x, B) \stackrel{\text{def}}{=} \inf\{\|x - x'\| : x' \in B\}, \tag{2.9}$$

in which $\|\cdot\|$ denotes the norm for $\mathbb{R}^n$. We observe that for a given smooth $\partial O$, the corresponding $\mathcal{P}_h$ may become non-smooth for some $h(\cdot) > 0$. Consider again Example 2.1. Evidently, for $h < r_c$, the radius of curvature of the middle section of $\mathcal{O}$, the constant height observation platforms $\mathcal{P}_h$ are smooth. When $h = r_c$, $\mathcal{P}_h$ has a kink or cusp at the circle $x_1^2 + x_3^2 = x_{3o}^2$. For $h > r_c$, there exist more than one point in $\partial\mathcal{O}$ that are at height $h$ from the observation platform $\mathcal{P}_h$. In this case, the height of an observation point $z \in \mathcal{P}_h$ from $\partial\mathcal{O}$ is defined as $h = \inf\{\|z - y\| : y \in \partial\mathcal{O}\}$. Figure 2.20 shows the cross-section of $\mathcal{P}_h$ in the $x_1 = 0$ plane for Example 2.3 with $x_{2o} = 5, x_{3o} = 8, r_o = 4$ and various values of $h > 0$. In this example, there exist two or more distinct points $x^{(i)} \in \partial\mathcal{O}$ such that the outward rays $\mathcal{R}(x^{(i)}) = \{x \in \mathbb{R}^3 : x = x^{(i)} + \lambda n(x^{(i)}), \lambda \geq 0\}$ along the outward normal $n(x^{(i)})$ intersect at a point $\hat{x}$ corresponding to $h = r_c$, and this point is equidistant from the points $x^{(i)}$. This property motivates the following definition (see Fig. 2.21):

**Definition 2.3** A point $\hat{x} \in \mathbb{R}^3$ is an *outward-normal intersection point* of a surface $\partial\mathcal{O}$, if there exist two or more distinct points $x^{(i)} \in \partial\mathcal{O}$ such that the rays $\mathcal{R}(x^{(i)}) = \{x \in \mathbb{R}^3 : x = x^{(i)} + hn(x^{(i)}), h \geq 0\}$ along the outward unit normal $n(x^{(i)})$ intersect at some point $\hat{x}$.

*Remark 2.6* In the above definition, the values of $h$ corresponding to the rays at the intersection point $\hat{x}$ may not be the same. In the special case where $h$ is independent of $x \in \partial\mathcal{O}$, the outward-normal intersection point $\hat{x}$ of a surface is equidistant from

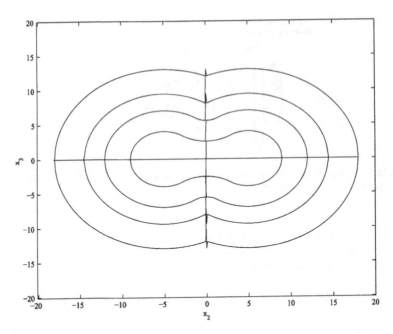

**Fig. 2.20** Cross-sections of $\mathcal{P}_h$ in Example 2.3 with $x_{2o} = 5$, $x_{3o} = 8$, $r_o = 4$ for $h = 0$ (*innermost curve*), 3, $h = r_c = \sqrt{89} - 4$ and $h = 9$ (*outermost curve*)

**Fig. 2.21** A 3-D object with an outward-normal intersection point

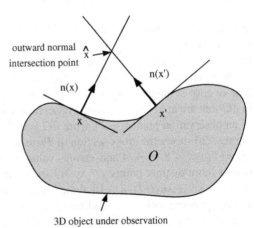

outward normal intersection point

$\hat{x}$

$n(x')$

$n(x)$

$x$

$x'$

$O$

3D object under observation

the points $x^{(i)} \in \partial \mathcal{O}$ for some constant $h > 0$. Then $\hat{x}$ corresponds to the usual *focal point* of a surface in differential geometry [4]. Note that in geometric optics, the lengths of the rays connecting the focal point and any point $x \in \partial \mathcal{O}$ are equal so that the optical signals emanating from all $x \in \partial \mathcal{O}$ arrived at the focal point are in phase. This requirement is not imposed in the definition of an outward-normal intersection point.

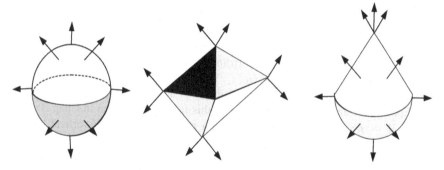

**Fig. 2.22** 3-D objects with no outward-normal intersection points

Typical 3-D objects in $\mathbb{R}^3$ having no outward-normal intersection points are convex bodies whose boundaries are spheroids, ellipsoids, zonoids and convex polytopes (see Fig. 2.22).

*Remark 2.7* For Cases (i)(a) with $\mathcal{W} = \mathbb{R}^2$ or (ii)(a) with $\mathcal{W} = \mathbb{R}^3$ in Sect. 2.1, the object under observation is respectively a plane curve or a surface corresponding to the graph $G_f$ of a smooth function $f = f(x)$ defined on a compact subset of $\mathcal{W}$, and the observation platform $\mathcal{P}$ is at a constant *vertical-height* $h_v$ above the plane curve or surface specified by the graph of $f_{h_v} \stackrel{\text{def}}{=} f + h_v$. Note the distinction between $h$ and $h_v$. They denote respectively the height along the surface normal $n(x)$ and vertical-height. Clearly, $\mathcal{P}$ corresponds to the elevated $G_f$ which has the same smoothness property as that of $G_f$ for all $h_v > 0$. Thus, the change in smoothness of $\mathcal{P}$ with respect to $h_v$ as discussed here does not occur.

The following prepositions pertaining the smoothness of the observation platform $\mathcal{P}_h$ are evident from Definition 2.3.

**Proposition 2.4** *If the surface $\partial\mathcal{O}$ of a compact solid body $\mathcal{O} \subset \mathbb{R}^3$ is smooth, and has no outward-normal intersection points, then the observation platform $\mathcal{P}_h$ and $\partial\mathcal{O}$ have the same smoothness for any $h(\cdot) > 0$.*

**Proposition 2.5** *If the surface $\partial\mathcal{O}$ of a compact solid body $\mathcal{O} \subset \mathbb{R}^3$ is smooth, and has a finite number of outward-normal intersection points $\hat{x}^{(i)}$ with corresponding heights $h_i, i = 1, \ldots, M$, along the outward normal $n(x^{(i)})$ above $x^{(i)}$, then any observation platform $\mathcal{P}_h$ with constant $h < \min\{h_i, i = 1, \ldots, M\}$ and $\partial\mathcal{O}$ have the same smoothness.*

In practical applications, it is of interest to consider the sensitivity and robustness of total visibility with respect to positional perturbations associated with a given set of point-observers. For preciseness, we introduce the following definitions:

**Definition 2.4** Given a single point-observer at $z_o \in \mathcal{P}$ from which the observed-object surface $\partial\mathcal{O}$ is totally visible, $z_o$ is said to be *robust with respect to position*

*perturbations*, if there exists an $\epsilon$-neighborhood of $z_o$ such that total-visibility is invariant, i.e. for each $z' \in (\{z \in \mathcal{W} : \|z - z_o\| \leq \epsilon\} \cap \mathcal{P})$, its visible set $\mathcal{V}(z') = \partial \mathcal{O}$.

For a simple example, consider the case where the object $\mathcal{O}$ is a smooth surface $G_f \subset \mathbb{R}^3$ as described by Case (ii)(a) in Sect. 2.1, with a constant vertical-height observation platform $\mathcal{P}_{h_v} = G_{f+h_v}$. We mention here that there exists a critical vertical-height $h_{vc}$ such that total visibility is attained for any point-observer $z_o \in \text{Epi}_{f+h_v}$ for any $h_v > h_{vc}$ (A proof of this fact will be given in Chap. 3), i.e. $\mu_2\{\mathcal{V}(z_o)\} = \mu_2\{G_f\}$. Thus, any observation point $z \in \mathcal{P}_{h_v}$ with vertical height $h_v > h_{vc}$ is robust with respect to position perturbations.

The sensitivity of the visible-set measure of a point-observer with respect to position perturbations can be defined in terms of the Euclidean norm of the gradient of the visible-set measure as follows:

**Definition 2.5** The sensitivity of the visible-set measure of a point-observer at a given observation point $z_o$ with respect to position perturbations is defined as

$$\sigma(z_o) = \|\nabla_z \mu_2\{\mathcal{V}(z)\}|_{z=z_o}\|. \tag{2.10}$$

Evidently, for the foregoing simple example, $\sigma(z_o) = 0$ for any $z_o \in \mathcal{P}_{h_v}$ with $h_v > h_{vc}$.

## 2.3 Observation of Complex Objects

So far, we have considered only the observation of an object consisting of a single connected compact set in the world space $\mathcal{W}$. In more general situations, the object $\mathcal{O}$ under observation may correspond to a collection of *disjoint* compact subsets $\mathcal{O}_i, i \in \mathcal{I}$ of $\mathcal{W}$, where $\mathcal{I}$ is a finite or countably infinite index set. The observations of $\mathcal{O}$ are made from points $z \in \mathcal{P} \subset \mathcal{W}$ such that $\mathcal{P} \cap \mathcal{O}$ is empty. In the trivial case where the observation of any object $\mathcal{O}_i$ can be made from a given observation point $z \in \mathcal{P}$ without considering the remaining subsets $\mathcal{O}_j, j \in \mathcal{I} - \{i\}$, then the visible set $\mathcal{V}(z)$ is simply $\bigcup_{i \in \mathcal{I}} \mathcal{V}_i(z)$, where $\mathcal{V}_i(z) \subset \partial \mathcal{O}_i$ denotes the visible set of $z$ with respect to $\mathcal{O}_i$. To illustrate various possible situations involving objects with multiple disjoint subsets, consider a simple example where the object $\mathcal{O} \subset \mathbb{R}^2$ consists of two circular disks $D_i, i = 1, 2$ with different radii $r_1$ and $r_2$ as shown in Fig. 2.23. First, consider the observation point $z^{(1)}$ located between the two disks. Evidently, both $D_1$ and $D_2$ are partially visible from $z^{(1)}$. From the observation point $z^{(2)}$ (resp. $z^{(3)}$), only $D_1$ (resp. $D_2$) is partially visible (as in solar eclipse, where the observation point $z$ is identified with the sun). From the observation $z^{(4)}$, both $D_1$ and $D_2$ are partially visible. However, only one point in $D_1$ is visible from $z^{(4)}$. Finally, both $D_1$ and $D_2$ are partially visible from the observation point $z^{(5)}$. Moreover, $\mathcal{V}(z^{(5)})$, the visible set of $z^{(5)}$ is simply $\mathcal{V}_1(z^{(5)}) \cup \mathcal{V}_2(z^{(5)})$, where $\mathcal{V}_i(z^{(5)})$ denotes the visible set of $z^{(5)}$ with respect to $D_i$, which can be determined independently. This simple example shows

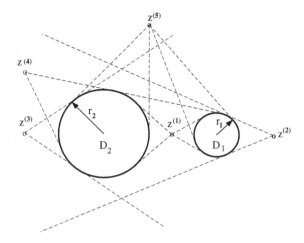

**Fig. 2.23** Object composed of two disjoint circular disks in $\mathbb{R}^2$

that the structure of the visible sets for objects with multiple disjoint subsets may be very complex, and their determination may be a computationally intensive task.

The following is an example of an object in $\mathcal{W} = \mathbb{R}^3$ : composed of an countably infinite number of disjoint compact subsets:

*Example 2.4* Let the object under observation be $\mathcal{O} = \cup_{i \in \mathcal{I}} \mathcal{O}_i$, where $\mathcal{I} = \{1, 2, \ldots\}$, and $\mathcal{O}_i$ is a circle with radius $r_i$, centered at $(x_1, x_2, x_3) = (0, 0, x_{3i})$:

$$\mathcal{O}_i = \{x = (x_1, x_2, x_3) \in \mathbb{R}^3 : x_1^2 + x_2^2 = r_i^2, x_3 = x_{3i} = 1/i^2\}, \quad i = 1, 2, \ldots. \tag{2.11}$$

such that $\mathcal{O}_i \cap \mathcal{O}_j$ is empty for all $i, j \in \mathcal{I}, i \neq j$. The observation platform $\mathcal{P}$ may correspond to a nonempty compact subset of $\mathcal{O}^c$, the complement of $\mathcal{O}$ relative to $\mathcal{W}$. The visible set of an observation point $z \in \mathcal{P}$ is given by

$$\mathcal{V}(z) = \bigcup_{i \in \mathcal{I}} \mathcal{V}_i(z), \tag{2.12}$$

where $\mathcal{V}_i(z)$ is the set of all boundary points of $\mathcal{O}_i$ that is visible from $z$. The object $\mathcal{O}$ is said to be totally visible from $z$ if $\mathcal{V}(z) = \bigcup_{i \in \mathcal{I}} \partial \mathcal{O}_i$, i.e. the boundary points of every $\mathcal{O}_i$ are visible from $z$. Evidently, there does not exist an observation point $z \in \mathcal{O}^c$ from which $\mathcal{O}$ is totally visible.

In planetary explorations using mobile robots, one may encounter cavities and tunnel-like structures on the planet surface. Here, the object $\mathcal{O}$ under observation is the surface inside these structures. To describe $\mathcal{O}$ mathematically, consider the simple idealized case in the world space $\mathcal{W} = \mathbb{R}^2$ where $\mathcal{O}$ is the union of the graphs of two real-valued $C_1$ functions $f_i = f_i(x), i = 1, 2$ defined on the interval $[a, b]$ as shown in Fig. 2.24. In Fig. 2.24a, $G_{f_1}$ and $G_{f_2}$ intersect at $x = b$ where $f_1(b) = f_2(b)$. Thus,

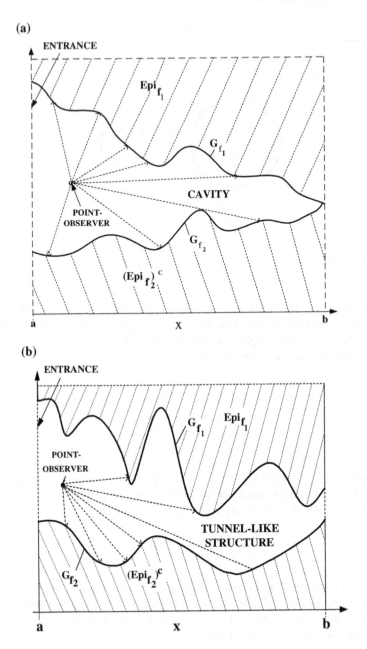

**Fig. 2.24** **a** Cavity and **b** tunnel-like structures

the region corresponding to $(\mathrm{Epi}_{f_1})^c \cap \mathrm{Epi}_{f_2}$ is an one-dimensional cavity in which the point-observers lie. When $G_{f_1}$ and $G_{f_2}$ do not intersect, we have a tunnel-like structure as shown in Fig. 2.24b. Now, given $f_1$ and $f_2$, it is of interest to determine whether there exists a point-observer $z$ in the interior of $(\text{Epi}_{f_1})^c \cap \text{Epi}_{f_2}$ such that $\mathcal{O}$ is totally visible from $z$. Evidently, if the given $f_1$ and $f_2$ generate a convex region $(\text{Epi}_{f_1})^c \cap \text{Epi}_{f_2}$, then $\mathcal{O}$ is totally visible from any point-observer in the interior of $(\text{Epi}_{f_1})^c \cap \text{Epi}_{f_2}$. Actually, total-visibility is attainable by a single point-observer even $(\text{Epi}_{f_1})^c \cap \text{Epi}_{f_2}$ is non-convex as demonstrated by the following simple example:

*Example 2.5* Let $f_1$ and $f_2$ be real-valued functions defined on $\Omega = [-1, 1]$ described by

$$f_1(x) = (1 + x^2); \quad f_2(x) = -f_1(x). \tag{2.13}$$

The object under observation is given by $\mathcal{O} = G_{f_1} \cup G_{f_2}$ as shown in Fig. 2.25. It is evident that $(\text{Epi}_{f_1})^c \cap (\text{Epi}_{f_2}) = \{(x, y) \in \mathbb{R}^2 : |y| \le f_1(x); x \in \Omega\}$ is non-convex, and $\mathcal{O}$ is totally visible from a single point-observer at the origin.

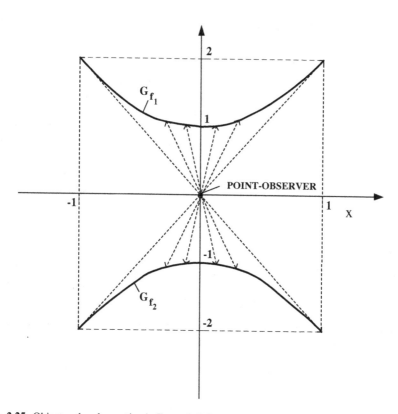

**Fig. 2.25** Object under observation in Example 2.5

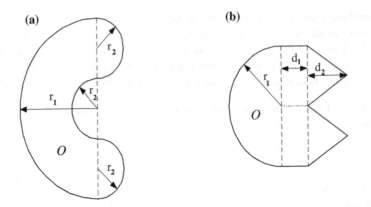

**Fig. 2.26** Objects with boundaries composed of **a** circular arcs and **b** both straight-line segments and circular arcs

## 2.4 Concluding Remarks

In this chapter, we have introduced various notions of visibility associated with line-of-sight observation of an object from a point-observer. These notions are also applicable to target interception by replacing the point-observer with a point source, and the object under observation with a target (e.g. a laser source emitting a beam toward the target). We may rename the "visible set of a point observer" as the "impact set of the point source". Thus, total visibility of an object from a point-observer corresponds to total impact of the target boundary by a point source, i.e. each point of the target boundary can be impacted by at least one straight beam emitted from the source. In practical applications, it may be of interest to expose a particular part of the target surface to the source. This task can be accomplished only when that particular part lies in the impact set of the source.

### Exercises

**Ex.2.1.** Let $\mathcal{W} = \mathbb{R}^2$. For each of the objects $\mathcal{O} \subset \mathcal{W}$ whose boundaries are composed of circular arcs and straightline segments (see Fig. 2.26), determine the smallest number and locations of point-observers in $\mathcal{O}^c$ for total visibility of $\mathcal{O}$.

**Ex.2.2.** Let $\mathcal{W} = \mathbb{R}^2$. The object $\mathcal{O}$ under observation is formed by the union of two circular disks with different radii $r_1, r_2 > 0$.

(i) Find the visible sets of various observation points $z \in \mathcal{O}^c$ for each of the following cases:

  (a) two disks are tangent to each other;
  (b) two disks are disjoint with their centers separated by a finite distance $r_1 + r_2 + d$, $d > 0$.

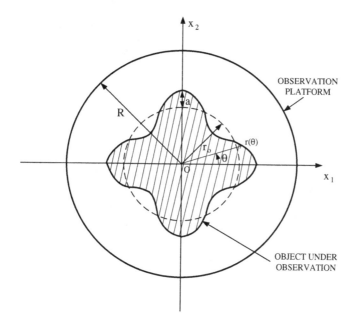

**Fig. 2.27** A circular disk with rippled boundary $n = 4$

(ii) Determine the constant-height observation platforms $\mathcal{P}_h$ for cases (a) and (b) in (i). For each case, examine the smoothness of $\mathcal{P}_h$ as $h$ increases from 0.

**Ex.2.3.** Let the object under observation $\mathcal{O}$ be a circular disk in $\mathcal{W} = \mathbb{R}^2$ with mean radius $r_o$ and rippled boundary described by $r(\theta) = r_o + a\cos(n\theta)$, $0 \leq \theta \leq 2\pi$, $n = 1, 2, \ldots$ (see Fig. 2.27), where $a$ is a positive number $< r_o$. The observation platform $\mathcal{P}$ is a circle centered at the origin with radius $R > r_o + a$.

(a) Given any $n = 1, 2, \ldots$, determine the smallest number and locations of point-observers in $\mathcal{P}$ for total visibility of $\mathcal{O}$.
(b) Discuss the solution to the problem in (a) as $n \to \infty$.

**Ex.2.4.** Extend the notion of a radially-deformed circular disk in $\mathbb{R}^2$ to a radially-deformed sphere in $\mathbb{R}^3$, and develop results pertaining to total visibility using a finite number of point-observers located at finite distances from the radially-deformed sphere.

**Ex.2.5.** Let $\mathcal{W} = \mathbb{R}^2$. The object $\mathcal{O}$ under observation is composed of three identical circular disks $D_i$, $i = 1, 2, 3$ with radius $r_o$, whose centers are at the vertices of an equilateral triangle centered at the origin as shown in Fig. 2.28. The observation platform $\mathcal{P}$ is a circle centered at the origin with radius $R > r_o$. Partition $\mathcal{P}$ into subset sets $\mathcal{P}_j$ from which certain subsets of $\{D_i, i = 1, 2, 3\}$ are partially visible. Is $\mathcal{O}$ totally visible from $\mathcal{P}$.

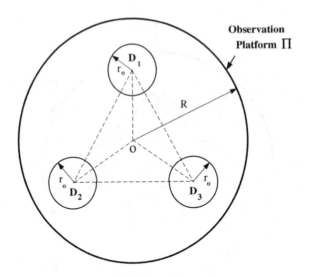

**Fig. 2.28** Object under observation for Exercise 2.5

**Ex.2.6.** Let $\mathcal{W} = \mathbb{R}^3$. The object $\mathcal{O}$ under observation is a rocket engine composed of a spherical ball with radius $R$ attached to a truncated cone with length $L$ and terminal radii $r_1, r_2$ as shown in Fig. 2.29. Find the visible sets corresponding to various observation points $z \in \mathcal{O}^c$.

**Ex.2.7.** Let $\mathcal{W} = \mathbb{R}^3$. The object $\mathcal{O}$ under observation is a surface described by $\mathcal{O}_1 \cup \mathcal{O}_2$, where

$$\mathcal{O}_1 = \{(x_1, x_2, x_3) \in \mathbb{R}^3 : (x_1 + 1)^2 + x_2^2 \le 1, x_3 = (x_1 + 1)^2 + x_2^2)\};$$
$$\mathcal{O}_2 = \{(x_1, x_2, x_3) \in \mathbb{R}^3 : (x_1 - 1)^2 + x_2^2 \le 1, x_3 = (x_1 - 1)^2 + x_2^2)\}. \quad (2.14)$$

The observations are made from the constant vertical-height platform $\mathcal{P}_{h_v}$ defined by

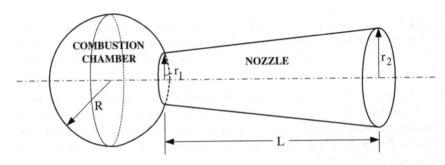

**Fig. 2.29** Rocket engine under observation in Exercise 2.6

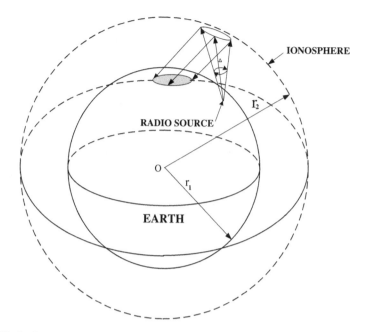

**Fig. 2.30** Radio-wave reflection from ionosphere

$$\mathcal{P}_{h_v} = \{(x_1, x_2, x_3) \in \mathbb{R}^3 : x_1 = x_1', x_2 = x_2', x_3 = x_3' + h_v, (x_1', x_2', x_3') \in \mathcal{O}\}, \quad h_v > 0.$$
(2.15)

(a) For a given positive vertical height $h_v \leq 1$, determine the visible sets for various observation points $z \in \mathcal{P}_{h_v}$.
(b) Repeat (a) for vertical height $h_v > 1$.

**Ex.2.8.** A radio point-source located on the surface of the spherical Earth with radius $r_1$ emits radio beam with solid angle $\Delta$. The beam is reflected from the ionosphere (a reflective spherical shell with radius $r_2$) back to Earth (see Fig. 2.30). The reflection obeys the usual law of angle of incidence equals the angle of reflection. Describe in mathematical terms the Earth's surface area that is exposed to the reflected radio beam.

**Ex.2.9.** Consider the objects $\mathcal{O}$ under observation as shown in Fig. 2.26. Construct an example for each case (a) and case (b) by choosing appropriate $C_1$-functions $f_1 = f_1(x)$ and $f_2 = f_2(x)$ such that total visibility of $\mathcal{O} = G_{f_1} \cup G_{f_2}$ requires at least two point-observers in the interior of the region $(\text{Epi}_{f_1})^c \cap \text{Epi}_{f_2}$. Give the coordinates of these point-observers.

**Ex.2.10.** Determine whether the following statement is (i) true, (ii) false, or (iii) sometimes true and sometimes false? In (iii), give a simple example for each case.

If the convex hull of a nonempty compact subset $S$ of $\mathbb{R}^2$ (the smallest convex set containing $S$) has interior points, then the boundary of $S$ is totally visible from any interior point of the convex hull of $S$.

# References

1. P.K.C. Wang, A class of optimization problems involving set measures. Nonlinear Anal. Theor. Methods Appl. **47**, 25–36 (2001)
2. P.K.C. Wang, Optimal path planning based on visibility. J. Optim. Theor. Appl. **117**, 157–181 (2003)
3. D. Luenberger, *Linear Vector Spaces and Optimization* (Wiley, New York, 1979)
4. J.A. Thorpe, *Elementary Topics in Differential Geometry* (Springer, New York, 1979)

# Chapter 3
# Static Optimal Visibility Problems

Having introduced in Chap. 2 the notions of total and partial visibility of an object from point-observers attached to a given observation platform, we can now proceed to formulate various optimal visibility problems. In this chapter, we shall focus our attention on static optimal visibility problems in which the observers are stationary with respect to the object under observation. To fix ideas and reveal some of the intrinsic difficulties of these problems, we shall begin with the simplest case of a single point-observer. Then, static optimal visibility problems involving multiple point-observers will be discussed.

## 3.1 Single Point-Observer Static Optimal Visibility Problems [1]

Consider the simplest case where the observed object $\mathcal{O}$ and the observation platform $\mathcal{P}$ are respectively the graphs of specified real-valued $C_1$-functions $f = f(x)$ and $g = g(x)$ defined on $\Omega$, a simply connected, compact subset of $\mathbb{R}^n$, $n \in \{1, 2\}$ such that

$$g(x) > f(x) \quad \text{for all } x \in \Omega. \tag{3.1}$$

As mentioned in Remark 2.3, a special observation platform having practical importance is the constant vertical-height platform corresponding to the *elevated profile of* $f$ defined by the graph of $f_{h_v} \overset{\text{def}}{=} f + h_v$, where $h_v$ is a given positive number specifying the vertical-height of the point-observer above $\mathcal{O} = G_f \overset{\text{def}}{=} \{(x, f(x)) \in \mathbb{R}^{n+1} : x \in \Omega\}$. Since $f$ is a $C_1$-function defined on a compact set $\Omega$, $G_f$ is also compact. Moreover, for any point-observer at $(x, g(x)) \in G_g$, its visible set $\mathcal{V}((x, g(x)))$ and its projection on $\Omega$ (denoted by $\Pi_\Omega \mathcal{V}((x, g(x)))$) are compact. Thus, we may regard $(x, g(x)) \to \mathcal{V}((x, g(x)))$ (resp. $\Pi_\Omega \mathcal{V}((x, g(x)))$) as a set-valued mapping on $G_g$ into $2^{G_f}$ (resp. $2^\Omega$). In general, $\mathcal{V}((x, g(x)))$ and $\Pi_\Omega(\mathcal{V}((x, g(x))))$ may be the union of *disjoint* compact subsets of $G_f$ and $\Omega$ respectively. This situation is illustrated by the example shown in Fig. 3.1 with the point-observer at $(x_o, g(x_o)) \in G_g$ and

© Springer International Publishing Switzerland 2015
P.K.-C. Wang, *Visibility-based Optimal Path and Motion Planning*,
Studies in Computational Intelligence 568, DOI 10.1007/978-3-319-09779-4_3

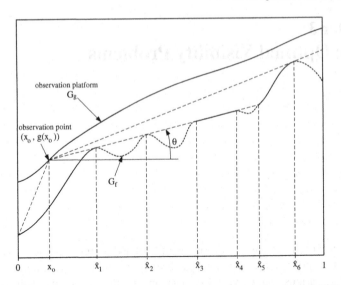

**Fig. 3.1** Example illustrating $\Pi_\Omega \mathcal{V}((x_o, g(x_o)))$ is the union of disjoint subsets of [0,1]

$\Omega = [0, 1]$. It can be seen that $\Pi_\Omega \mathcal{V}((x_o, g(x_o))) = [0, \hat{x}_1] \cup \{\hat{x}_2\} \cup [\hat{x}_3, \hat{x}_4] \cup [\hat{x}_5, \hat{x}_6]$. As in Example 2.1, this example also shows that the visible set of a point-observer may contain isolated points.

Now, we consider two optimal visibility problems associated with observation of the object $\mathcal{O} = G_f$ from point-observers located in Epi$_f$, the epigraph of $f$.

**Problem 3.1  Minimum Vertical-height Total Visibility Problem.** Given $f = f(x)$ defined on $\Omega$, find the minimum vertical-height $h_v^* \geq 0$ and a point $x^* \in \Omega$ such that $G_f$ is totally visible from the point-observer at $(x^*, f_{h_v^*}(x^*))$.

**Problem 3.2  Maximum Visibility Problem.** Given real-valued $C_1$-functions $f$ and $g$ defined on $\Omega$ satisfying condition (3.1), find a point $x^* \in \Omega$ such that $J_g(x^*) \geq J_g(x)$ for all $x \in \Omega$, where $J_g(x) \overset{\text{def}}{=} \mu_1\{\Pi_\Omega \mathcal{V}((x, g(x)))\}$, the Lebesgue measure of $\Pi_\Omega \mathcal{V}((x, g(x)))$.

If we set $g$ in Problem 3.2 to $f_{h_v}$ for a given $h_v > 0$, then we have the practically important *Constant Vertical-height Maximum Visibility Problem*.

*Remark 3.1* In Problem 3.2, we may choose to maximize the total measure of $\mathcal{V}((x, g(x)))$ instead of the total measure of $\Pi_\Omega \mathcal{V}((x, g(x)))$ at the expense of increased computational complexity.

To fix ideas, we first consider the foregoing problems for the case with an one-dimensional domain $\Omega$ and present some results which are relevant to the solution of more general optimal visibility problems. Then, similar problems for the case of a 2-dimensional $\Omega$ will be discussed.

**Case (i) Dim** $(\Omega) = 1$ : Let $\Omega$ be the compact interval $I = [a, b] \subset \mathbb{R}$, and $f = f(x)$ a given real-valued $C_1$-function defined on $I$. To solve Problems 3.1 and 3.2, we introduce a few preliminary results.

**Lemma 3.1** *For each point $x \in I$, there exists a critical vertical-height $h_{vc}(x) \geq 0$ such that $\mathcal{V}((x, f(x) + h_{vc}(x))) = G_f$, or $G_f$ is totally visible from the point-observer at $(x, f(x) + h_{vc}(x))$.*

*Proof* For the trivial case where $f$ is a convex $C_1$-function on $I$, $h_{vc}(x) = 0$ for any $x \in I$. This is also valid for any convex function $f$ having a piecewise continuous derivative on $I$. For nontrivial cases, we shall establish the existence of a $h_{vc} = h_{vc}(x)$ by construction. First, let $x$ be an interior point of $I$. We partition $I$ into two subintervals $I^-(x) = [a, x]$ and $I^+(x) = [x, b]$. Since $f$ is $C_1$ on $I$, $Df$ (the first derivative of $f$) has a maximum $Df_{max}^-(x)$ on $I^-(x)$, and a minimum $Df_{min}^+(x)$ on $I^+(x)$. Let $S^-(x) = \{x^- \in I^-(x) : Df(x^-) \geq 0\}$. Consider $(x, f(x^-) + h_{vc}^-(x, x^-))$ corresponding to the point of intersection between the vertical line passing through $x$ with the tangent line $L^-$ at $(x^-, f(x^-))$ (see Fig. 3.2), where

$$h_{vc}^-(x, x^-) = Df(x^-)(x - x^-). \tag{3.2}$$

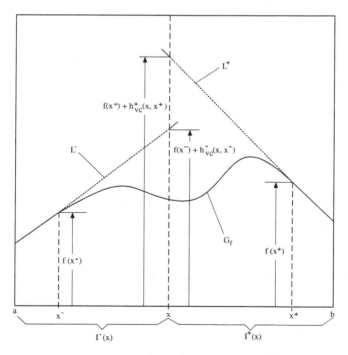

**Fig. 3.2** Construction of $h_{vc}^-(x, x^-)$ and $h_{vc}^+(x, x^+)$

Evidently, $|h_{vc}^-(x, x^-)| < \infty$, and the point $(x^-, f(x^-))$ is visible from $(x, f(x^-) + h_{vc}^-(x, x^-))$. Moreover, the set $H^-(x) = \{h_{vc}^-(x, x^-) : x^- \in S^-(x)\}$ is compact. Let $\hat{h}_{vc}^-(x)$ be the maximum element in $H^-(x)$. Then, $G_f$ restricted to $I^-(x)$ is visible from $(x, f(x) + \hat{h}_{vc}^-(x))$.

Similarly, let $S^+ = \{x^+ \in I^+(x) : Df(x^+) \leq 0\}$. Consider $(x, f(x^+) + h_{vc}^+(x, x^+))$ corresponding to the point of intersection between the vertical line passing through $x$ with the tangent line $L^+$ at $(x^+, f(x^+))$ (see Fig. 3.2), where

$$h_{vc}^+(x, x^+) = Df(x^+)(x - x^+). \tag{3.3}$$

Let $\hat{h}_{vc}^+(x)$ be the minimum element in $H^+(x) = \{h_{vc}^+(x, x^+) : x^+ \in S^+(x)\}$. Then, $G_f$ restricted to $I^+(x)$ is visible from the point $(x, f(x^+) + \hat{h}_{vc}^+(x))$. Now, we set $h_{vc}(x) = \max\{\hat{h}_{vc}^-(x), \hat{h}_{vc}^+(x)\}$. Thus, $G_f$ is visible from the point-observer at $(x, f(x) + h_{vc}(x))$. Finally, for $x = a$ or $b$, we set $h_{vc}(x) = \hat{h}_{vc}^+(x)$ or $\hat{h}_{vc}^-(x)$ respectively. Clearly, $G_f$ is not totally visible from any point-observer at $(x, f(x) + h_v(x))$ with $h_v(x) < h_{vc}(x)$. Hence, $h_{vc}(x)$ is the critical vertical-height at $x$.    □

*Remark 3.2* From the proof of Lemma 3.1, it is evident that if $Df_{\max}^-(x) > 0$ and $Df_{\min}^+(x) < 0$ for all $x \in I$, then the *critical vertical-height profile* defined by $h_{vc} = h_{vc}(x)$, $x \in I$, and its corresponding graph $G_{f+h_{vc}}$ is a convex curve formed by straightline segments described by (3.2) and (3.3). Moreover, the epigraph of $f + h_{vc}$ is the intersection of the epigraphs of all $(f + h_{vc}^-)$'s and $(f + h_{vc}^+)$'s.

*Remark 3.3* Lemma 3.1 remains valid for any real-valued function $f$ whose derivative $Df$ is a piecewise continuous function on $I$.

The next result follows from the fact that if $\tilde{h}_v(x) > h_v(x)$, then $\mathcal{V}((x, f(x) + h_v(x))) \subseteq \mathcal{V}((x, f(x) + \tilde{h}_v(x)))$, and $\mathcal{V}((x, f(x) + h_{vc}(x))) = G_f$.

**Lemma 3.2** *For any fixed* $x \in I$, $\mu_1\{\Pi_I \mathcal{V}((x, f(x) + h_v(x)))\}$ *increases monotonically for* $0 \leq h_v(x) \leq h_{vc}(x)$, *and* $\mu_1\{\Pi_I \mathcal{V}((x, f(x) + h_{vc}(x)))\}$ *(the measure of* $\Pi_I \mathcal{V}((x, f(x) + h_{vc}(x)))$*) is equal to* $(b - a)$.

Now, the solution to Problem 3.1, follows directly from Lemma 3.1.

**Proposition 3.1** *The set of all solutions to Problem 3.1 is given by* $G_{f_{h_v^*}} \cap G_{h_{vc}}$, *where* $h_v^* = \inf\{h_v : G_{f_{h_v}} \cap G_{h_{vc}} \neq \phi\}$.

The foregoing results are consistent with intuition that a smooth terrain in a bounded region is totally visible if the observation point is sufficiently high, and there exists a minimum vertical-height at which the entire terrain is visible.

Next, we consider Problem 3.2 in which the point-observers or observation points are restricted to the observation platform $\mathcal{P} = G_g$. Obviously, if $G_g \cap G_{h_{vc}}$ is nonempty, then there exists at least one solution $x^* \in I$ to this problem. Therefore we focus our attention on the nontrivial case where $G_g \cap G_{h_{vc}}$ is empty.

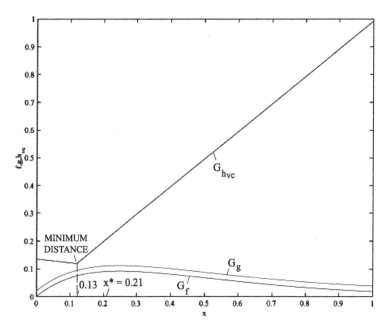

**Fig. 3.3** Counterexample for the statement that the point on $G_g$ closest to $G_{h_{vc}}$ gives maximum visibility of $G_f$

At first glance, one may conjecture that a solution $x^* \in I$ to Problem 3.2 could be characterized by

$$h_{vc}(x^*) - g(x^*) = \inf_{x \in I}(h_{vc}(x) - g(x)), \qquad (3.4)$$

implying that the point $(x^*, g(x^*))$ giving the maximum visibility of $G_f$ corresponds to the point on $G_g$ that is closest to $G_{h_{vc}}$ in the sense of vertical distance. A little reflection shows that this conjecture is true for some $f$ and $g$, but it is false in general. A counterexample is given by $f(x) = x \exp(-4x)$, $g(x) = f(x) + 0.02$ defined on $[0, 1]$. It can be verified that the minimum value of $(h_{vc}(x) - g(x))$ occurs at $x = 0.13$, and $x^* = 0.21$ (See Fig. 3.3).

Assuming that $G_g \cap G_{h_{vc}}$ is empty, consider a point $(x, g(x)) \in G_g$, and the ray $\mathcal{R}(\theta)$ emanating from this point with a specified angle $\theta$ with respect to the horizon (see Fig. 3.1). Let $(\hat{x}, g(\hat{x}))$ be a point of intersection of this ray with $G_f$. Then, the point $\hat{x}$ corresponds to a fixed point of the mapping of $w(\cdot; \theta, x)$ on $I$ into itself defined by

$$w(\hat{x}; \theta, x) = x - \cot(\theta)(g(x) - f(x)), \qquad (3.5)$$

if $\theta \neq 0$; and $\hat{x} = (f^{-1} \circ g)(x)$ if $\theta = 0$. Evidently, $(\hat{x}, f(\hat{x}))$ is either a point at which $\mathcal{R}(\theta)$ intersects $G_f$ at a nonzero angle (e.g. point $(\hat{x}_3, f(\hat{x}_3))$ in Fig. 3.1), or a point of tangency between $\mathcal{R}(\theta)$ and $G_f$ (e.g. points $(\hat{x}_i, g(x_i)), i = 1, 2, 3$

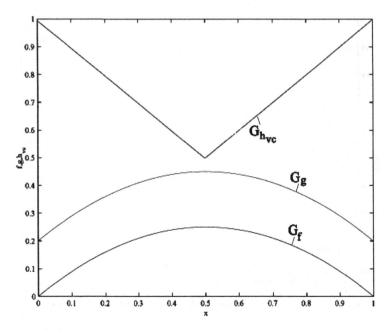

**Fig. 3.4** Graphs of $f$, $g = f_{h_v}$ and $h_{vc}$ for Example 3.1 with $h_v = 0.2$

in Fig. 3.1). In the first case, it can be readily verified that $\hat{x}$ is a continuous point of the mapping $\tilde{w}$ defined by $\tilde{w}(\hat{x}; x) \overset{\text{def}}{=} x + (f(\hat{x}) - g(x))/Df(\hat{x})$, provided that $Df(\hat{x}) \neq 0$. If $Df(\hat{x}) = 0$ then $\hat{x} = (f^{-1} \circ g)(x)$.

Consider again the example shown in Fig. 3.1 It is clear that for any $x \in [0, x_o]$, the sets $\{\hat{x}_2\}$ and $[\hat{x}_3, \hat{x}_4]$ are no longer visible from $(x, g(x))$. Consequently, the set-valued mapping $x \to \Pi_I \mathcal{V}((x, g(x)))$ is discontinuous at $x_o$ with respect to the usual metric on $I$, and the Hausdorff metric $\rho_H$ on $2^I$. It follows that $J_g = J_g(x)$, the measure of $\Pi_I \mathcal{V}((x, g(x)))$, is also discontinuous at $x_o$. Thus, Problem 3.2 may involve the maximization of a discontinuous real-valued function over a compact interval $I$. The following simple example exhibits this phenomenon.

*Example 3.1* Let $f(x) = x(1 - x)$ and $g(x) = f_{h_v}(x) \overset{\text{def}}{=} x(1 - x) + h_v, h_v > 0$, defined on $\Omega = [0, 1]$. The critical vertical-height profile $h_{vc}$ is given by $h_{vc}(x) = \max\{1 - x, x\}$. The graphs of $f$, $g$ and $h_{vc}$ are shown in Fig. 3.4. The visible set of any observation point $(x, g(x)) \in G_{f_{h_v}}(x)$ can be found by considering the ray emanating from a point $(x, f_{h_v}(x)) \in G_{f_{h_v}}$ that is tangent to $G_f$ at $(\hat{x}, f(\hat{x}))$ given by $Df(\hat{x}) = (f(\hat{x}) - f(x))/(\hat{x} - x)$. It can be verified that for $0 < h_v \leq 1/4$,

$$\Pi_I \mathcal{V}((x, f_{h_v}(x))) = \begin{cases} [0, x + \sqrt{h_v}] & \text{for } 0 \leq x \leq 1/2 - \sqrt{h_v}; \\ [x - \sqrt{h_v}, x + \sqrt{h_v}] & \text{for } 1/2 - \sqrt{h_v} \leq x \leq 1/2 + \sqrt{h_v}; \\ [x - \sqrt{h_v}, 1] & \text{for } 1/2 + \sqrt{h_v} \leq x \leq 1, \end{cases}$$

$$(3.6)$$

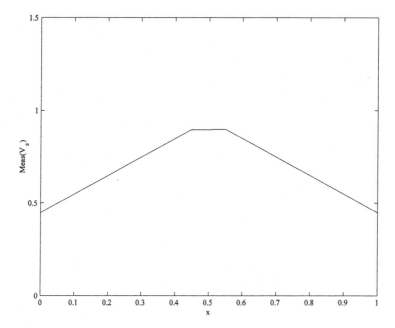

**Fig. 3.5**  Measure of visible sets for Example 3.1 with $h_v = 0.2$

and

$$J_{f_{h_v}}(x) \overset{\text{def}}{=} \mu_1\{\Pi_I \mathcal{V}((x, f_{h_v}(x)))\} = \begin{cases} x + \sqrt{h_v} & \text{for } 0 \le x \le 1/2; \\ 1 - x + \sqrt{h_v} & \text{for } 1/2 \le x \le 1. \end{cases} \quad (3.7)$$

Figure 3.5 shows the measure of the visible sets as a function of $x$ for $h_v = 0.2$. It is evident from the expression for $h_{vc}(x)$ that the solution to Problem 3.1 is given by $h_v^* = 1/4$. The solution to Problem 3.2 for $0 < h_v \le 1/4$ is given by the point set $\{1/2 - \sqrt{h_v}, 1/2 + \sqrt{h_v}\}$. This result shows that the set of optimal observation points does not always include the maximum points of $f$. Note that for $h_v = 1/4$, $J_{f_{h_v}}$ is continuous on $I$, but its derivative is discontinuous at $x^* = 1/2$.

For the special case where the mapping $x \to \Pi_I \mathcal{V}((x, g(x)))$ is locally Lipschitz continuous, we have the following result:

**Theorem 3.1**  *Let $I_i, i = 1, \ldots, N$, be open subintervals of $I$ such that $\bigcup_{i=1}^{N} \bar{I}_i = I$, where $\bar{I}_i$ denotes the closure of $I_i$. Suppose that for each $i \in \{1, \ldots, N\}$, there exists a positive constant $K_i$ such that*

$$\rho_H(\Pi_I \mathcal{V}((x', g(x'))), \Pi_I \mathcal{V}((x, g(x)))) \le K_i |x' - x| \quad \text{for all } x', x \in \bar{I}_i. \quad (3.8)$$

*Then, Problem 3.2 has a solution.*

*Proof*  Let $x, x' \in \bar{I}_i$; $S_x = \Pi_I \mathcal{V}((x, g(x)))$, and $S_{x'} = \Pi_I \mathcal{V}((x', g(x')))$. Consider

$$|J_g(x) - J_g(x')| \overset{\text{def}}{=} |\mu_1\{S_x\} - \mu_1\{S_{x'}\}|. \tag{3.9}$$

Since $S_x = \{S_x \cap S_{x'}^c\} \cup \{S_x \cap S_{x'}\}$ and $S_{x'} = \{S_x^c \cap S_{x'}\} \cup \{S_x \cap S_{x'}\}$, it follows that

$$|\mu_1\{S_x\} - \mu_1\{S_{x'}\}| = |\mu_1\{\Delta((x',x))\} - \mu_1\{\Delta((x,x'))\}| \le \mu_1\{\Delta((x',x))\} + \mu_1\{\Delta((x,x'))\}, \tag{3.10}$$

where $\Delta((x,x')) = S_x^c \cap S_{x'}$, and $\{\cdot\}^c$ denotes the complement of the set $\{\cdot\}$ relative to $I$.

The Hausdorff distance between the sets $S_x$ and $S_{x'}$ satisfies:

$$\begin{aligned}
\rho_H(S_x, S_{x'}) &\overset{\text{def}}{=} \max\{\max_{\tilde{x} \in S_x} \rho(\tilde{x}, S_{x'}), \max_{\tilde{x} \in S_{x'}} \rho(\tilde{x}, S_x)\} \\
&= \max\{\max_{\tilde{x} \in S_x \cap S_{x'}^c} \rho(\tilde{x}, S_{x'}), \max_{\tilde{x} \in S_{x'} \cap S_x^c} \rho(\tilde{x}, S_x)\} \\
&\ge \frac{1}{2}\left(\max_{\tilde{x} \in S_x \cap S_{x'}^c} \rho(\tilde{x}, S_{x'}) + \max_{\tilde{x} \in S_{x'} \cap S_x^c} \rho(\tilde{x}, S_x)\right),
\end{aligned} \tag{3.11}$$

where $\rho(\tilde{x}, S) \overset{\text{def}}{=} \min_{\tilde{x} \in S} |\tilde{x} - \bar{x}|$. Moreover, since $S_x$ and $S_{x'}$ are compact subsets of $I$, there exist positive constants $\kappa$ and $\kappa'$ such that

$$\mu_1\{\Delta((x,x'))\} \le \kappa \max_{\tilde{x} \in S_x \cap S_{x'}^c} \rho(\tilde{x}, S_{x'}), \quad \mu_1\{\Delta((x',x))\} \le \kappa' \max_{\tilde{x} \in S_{x'} \cap S_x^c} \rho(\tilde{x}, S_x). \tag{3.12}$$

It follows that

$$\mu_1\{\Delta((x',x))\} + \mu_1\{\Delta(x,x'))\} \le 2\max\{\kappa', \kappa\}\rho_H(\Pi_I V((x', g(x'))), \Pi_I V((x, g(x)))). \tag{3.13}$$

In view of (3.8), we have

$$\mu_1\{\Delta((x',x))\} + \mu_1\{\Delta((x,x'))\} \le 2\max\{\kappa', \kappa\}K_i|x' - x| \quad \text{for all } x', x \in \bar{I}_i. \tag{3.14}$$

Thus, $J_g$ is continuous on $\bar{I}_i$. From Weierstrass theorem for the existence of maximum of a real-valued continuous function defined on a compact set, $J_g$ has a maximum $J_g^{(i)}$ on $\bar{I}_i$. Moreover, $J_g$ has a maximum on $I$ given by $\max\{J_g^{(i)}, i = 1, \ldots, N\}$. $\square$

**Case (ii) Dim($\Omega$) = 2:** The basic ideas and results for Case (i) can be extended to the case where Dim($\Omega$) = 2. Here, $f$ is a real-valued $C_1$-function defined on $\Omega$, a specified simply connected compact subset of $\mathbb{R}^2$. To extend some of the notions and results for Case (i), we consider rays $\mathcal{R}(x; \eta) \overset{\text{def}}{=} \{x' \in \Omega : x' = x + \lambda\eta, \lambda \in \mathbb{R}\}$ emanating from any given point $x = (x_1, x_2) \in \Omega$ in the direction $\eta$, a unit vector in $\mathbb{R}^2$ as illustrated in Fig. 3.6. Let $f_{\mathcal{R}} = f_{\mathcal{R}}(\lambda; x, \eta)$ denote the real-valued function of one variable $\lambda$ corresponding to $f$ restricted to $\mathcal{R}(x; \eta)$. From Lemma 3.1, for each fixed $x \in \Omega$ and direction $\eta$, there exists a critical vertical-height $\tilde{h}_{vc}(x, \eta)$ such that

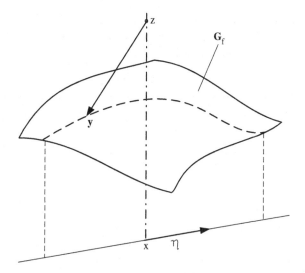

**Fig. 3.6** Computation of critical vertical-height for a surface

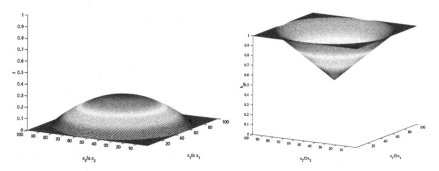

**Fig. 3.7** Observed surface and its critical vertical-height surface for Example 3.2; $\Delta x_1 = \Delta x_2 = 0.02$

$G_{f\mathcal{R}}$ is visible from the point $(x, \tilde{h}_{vc}(x, \eta))$. By considering $\tilde{h}_{vc}(x, \eta)$ for every unit vector $\eta \in \mathbb{R}^2$, we obtain the critical vertical-height at a point $x \in \Omega$ given by

$$h_{vc}(x) = \max_{\eta \in \mathbb{R}^2, \|\eta\|=1} \tilde{h}_{vc}(x, \eta). \qquad (3.15)$$

Thus, $h_{vc} = h_{vc}(x)$, $x \in \Omega$ defines a critical vertical-height surface, and Proposition 3.1 remains valid if we replace $I$ by $\Omega$, a compact, simply connected subset of $\mathbb{R}^2$.

*Example 3.2* Consider a 2-dimensional version of Example 3.1 with $f(r) = (1 - r^2)/4$ and $f_{h_v}(r) = (1 - r^2)/4 + h_v$ defined on the unit disk $\Omega = \{(r, \theta) : 0 \leq r \leq 1; 0 \leq \theta \leq 2\pi\}$. Here, the critical vertical-height profile is a conical surface given by $h_{vc} = (1 + r)/2$ defined on $\Omega$ (see Fig. 3.7).

Finally, Theorem 3.1 pertaining to the existence of solutions to Problem 3.2 can be extended to the case where $f$ is a real-valued $C_1$-function defined on a 2-dimensional $\Omega$, following the same arguments used in the proof.

*Remark 3.4* An optimization problem similar to the Maximum Visibility Problem 3.2 can be posed in which a cost density associated with the domain $\Omega$ is defined. We wish to minimize the total number of observation points and the total cost given by

$$\sum_{i=1}^{N} \int_{\mathcal{B}(x^{(i)};\delta)\cap\Omega} \gamma(x)d\Omega,$$

subject to the constraint specified by (1), where $\gamma$ is a specified nonnegative cost density function having compact support in $\Omega$; and $\mathcal{B}(x^{(i)};\delta)\overset{\text{def}}{=}\{x \in \mathbb{R}^2 : \|x - x^{(i)}\| < \delta\}$ for a given $\delta > 0$. This problem is applicable to the optimal placement of radio repeaters for cellular telephone networks with land cost.

Having discussed various optimal visibility problems associated with observation of one and two-dimensional objects, we turn our attention to similar problems associated with simply-connected three-dimensional objects.

**Lemma 3.3** *Suppose that the object under observation is a compact simply-connected solid body $\mathcal{O} \subset \mathbb{R}^3$ with a smooth boundary $\partial\mathcal{O}$ (at least $C_1$) and no outward-normal intersection points (see Definition 2.3). Then, total visibility of $\mathcal{O}$ can be attained by a finite set of observation points in the observation platform $\mathcal{P}_h$ at a finite height $h$ above the body surface $\partial\mathcal{O}$.*

*Proof* Since $\mathcal{O}$ is a compact simply-connected solid body with a smooth boundary $\partial\mathcal{O}$ and no outward-normal intersection points, then at any point $x \in \partial\mathcal{O}$, a unique outward normal $n(x)$ is defined. Moreover, the observation platform $\mathcal{P}_h = \{z \in \mathbb{R}^3 : z = x + hn(x), x \in \partial\mathcal{O}\}$ is well-defined. For any observation point $z \in \mathcal{P}_h$, there exists a point $x = z - hn(x) \in \partial\mathcal{O}$ such that $x$ is visible from $z$. Moreover, $\mathcal{V}(z)$, the visible set of $z$, is a compact subset of $\partial\mathcal{O}$ containing $x$ with finite Lebesgue measure or surface area $\mu_2(\mathcal{V}(z))$. Since $\mu_2(\partial\mathcal{O})$ is finite, there exists a finite number of observation points $z^{(i)} \in \mathcal{P}_h, i = 1, \ldots, N$ such that $\partial\mathcal{O} = \bigcup_{i=1}^{N}\mathcal{V}(z^{(i)})$ as guaranteed by the Heine-Borel Covering Theorem.                                       $\square$

In view of Lemma 3.3, a generalized version of Problem 3.2 for a single stationary observer can be formulated as follows:

**Problem 3.3 Single Stationary-Observer Maximum Visibility Problem.** Given an observation platform $\mathcal{P}$ enclosing the observed object $\mathcal{O} \subset \mathbb{R}^3$, find a point $z^* \in \mathcal{P}$ such that $J_0(z^*) \geq J_0(z)$ for all $z \in \mathcal{P}$, where $J_0(z) = \mu_2\{\mathcal{V}(z)\}$, the surface area of the visible set $\mathcal{V}(z)$.

As mentioned earlier, the set-valued mapping $z \to \mathcal{V}(z)$ is generally discontinuous with respect to the metric on $\mathcal{P}$ and the Hausdorff metric on $2^{\partial\mathcal{O}}$. Thus,

Problem 3.3 is generally a non-smooth optimization problem. A generalized version of Theorem 3.1 for the special case where the mapping $z \to \mathcal{V}(z)$ is Lipschitz continuous can be stated as follows:

**Theorem 3.2** *Let* $\mathcal{P}^{(i)}, i = 1, \ldots, N$, *be open subsets of the observation platform* $\mathcal{P}$ *such that* $\bigcup_{i=1}^{N} \bar{\mathcal{P}}^{(i)} = \mathcal{P}$, *where* $\bar{\mathcal{P}}^{(i)}$ *denotes the closure of* $\mathcal{P}^{(i)}$. *Suppose that for each* $i \in \{1, \ldots, N\}$, *there exists a positive constant* $K_i$ *such that*

$$\rho_H(\mathcal{V}(z), \mathcal{V}(z')) \leq K_i \rho_E(z, z') \text{for all } z, z' \in \bar{\mathcal{P}}^{(i)}, \tag{3.16}$$

*where* $\rho_H$ *and* $\rho_E$ *denote the Hausdorff and Euclidean metrics respectively. Then, Problem 3.3 has a solution.*

*Proof* Let $z, z' \in \bar{\mathcal{P}}^{(i)}$. Consider

$$|J_0(z) - J_0(z')| \overset{\text{def}}{=} |\mu_2\{\mathcal{V}(z)\} - \mu_2\{\mathcal{V}(z')\}| = |\mu_2\{\Delta((z, z'))\} - \mu_2\{\Delta((z', z))\}|$$

$$\leq |\mu_2\{\Delta((z, z'))\} + \mu_2\{\Delta((z', z))\}|,$$

where $\Delta((z, z')) = \mathcal{V}(z) \cap \mathcal{V}(z')^c$, and $\{\cdot\}^c$ denotes the complement of the set $\{\cdot\}$ relative to $\bar{\mathcal{P}}$. Now, the Hausdorff distance between $\mathcal{V}(z)$ and $\mathcal{V}(z')$ satisfies

$$\rho_H(\mathcal{V}(z), \mathcal{V}(z')) \overset{\text{def}}{=} \max\{\max_{\tilde{z} \in \mathcal{V}(z)} \rho(\tilde{z}, \mathcal{V}(z')), \max_{\tilde{z} \in \mathcal{V}(z')} \rho(\tilde{z}, \mathcal{V}(z))\}$$

$$= \max\{\max_{\tilde{z} \in \Delta((z, z'))} \rho(z, \mathcal{V}(z')), \max_{\tilde{z} \in \Delta((z', z))} \rho(z', \mathcal{V}(z))\}.$$

$$\geq \frac{1}{2}\left(\max_{\tilde{z} \in \Delta((z, z'))} \rho(z, \mathcal{V}(z')) + \max_{\tilde{z} \in \Delta((z', z))} \rho(z', \mathcal{V}(z))\right), \tag{3.17}$$

where $\rho(z, \mathcal{V}(z')) \overset{\text{def}}{=} \min_{\tilde{z} \in \mathcal{V}(z')} \rho_E(z, \tilde{z})$. Making use of (3.16) (Lipschitz continuity of $z \to \mathcal{V}(z)$) and following similar steps in the proof of Theorem 3.1 with obvious modifications lead to the desired result.   □

In general, a solution $z^* \in \mathcal{P}$ to Problem 3.3 may correspond to a point at which $J_0(\cdot) = \mu_2\{\mathcal{V}(\cdot)\}$ is not differentiable. Although necessary conditions for optimality can be derived in terms of quasi-differentials and generalized gradients of $J_0$ (see [2] and [3]), but useful efficient computational algorithms using these conditions remain to be developed.

## 3.2 Multiple Point-Observer Static Optimal Visibility Problems

So far, we have considered various optimal visibility problems involving a single stationary point-observer. When total visibility of the observed object cannot be achieved by a single stationary point-observer, it is natural to ask whether total visibility can be attained by a finite (preferably smallest) number of stationary point-observers. Before answering this question, we first establish a few properties of visible sets which are useful in the subsequent development. To simplify our discussion, we consider the case where the object $\mathcal{O}$ under observation is a surface in $\mathbb{R}^3$ described by $G_f$, the graph of a real-valued continuous function $f = f(x)$ defined on $\Omega$, a compact subset of $\mathbb{R}^2$. The observation points are restricted to a constant vertical-height observation platform $\mathcal{P}_{h_v} = G_{f+h_v}$.

**Lemma 3.4** *Every point $x' \in \Omega$ is a fixed-point of the set-valued mapping $x \rightarrow \Pi_\Omega \mathcal{V}((x, f_{h_v}(x)))$ on $\Omega$ into $2^\Omega$. Moreover, at a point $x' \in \Omega$ where the mapping $\Pi_\Omega \mathcal{V}(\cdot, f_{h_v}(\cdot))$ is continuous with respect to the Euclidean metric $\rho_E$ on $\Omega$, and Hausdorff metric $\rho_H$ on $2^\Omega$, there exists an open ball $\mathcal{B}(x'; \delta) = \{x \in \mathbb{R}^2 : \|x-x'\| < \delta\}$ about $x'$ with radius $\delta > 0$ such that $(\mathcal{B}(x'; \delta) \cap \Omega) \subset \Pi_\Omega \mathcal{V}((x', f_{h_v}(x')))$.*

*Proof* Let $x'$ be any point in $\Omega$. Then, the point $(x', f(x'))$ is always visible from the point $(x', f_{h_v}(x')) \in \text{Epi}_f$. Hence, $(x', f(x')) \in \mathcal{V}((x', f_{h_v}(x')))$, and $x' \in \Pi_\Omega \mathcal{V}((x', f_{h_v}(x')))$, or $x'$ is a fixed point of $\Pi_\Omega \mathcal{V}((\cdot, f_{h_v}(\cdot)))$. At a point $x' \in \Omega$ where the mapping $\Pi_\Omega \mathcal{V}(\cdot, f_{h_v}(\cdot))$ is continuous with respect to the metrics $\rho_E$ and $\rho_H$, there exists an open ball $\mathcal{B}(x'; \delta)$ with radius $\delta > 0$ such that for every $x \in \mathcal{B}(x'; \delta) \cap \Omega$, the point $(x, f(x))$ is visible from $(x, f_{h_v}(x))$. Thus, the desired result follows.                                                                                 $\square$

**Theorem 3.3** *Assume that the spatial domain $\Omega$ has a $C_1$-boundary $\partial\Omega$, and the mapping $x \rightarrow \Pi_\Omega \mathcal{V}((x, f_{h_v}(x)))$ from $\partial\Omega$ into $2^\Omega$ is continuous with respect to metrics $\rho_E$ and $\rho_H$. Then there exists an integer $N \geq 1$, and a finite point set $P^{(N)} = \{x^{(k)}, k = 1, \ldots, N\} \subset \Omega$ such that $\Omega = \bigcup_{k=1}^{N} \Pi_\Omega \mathcal{V}((x^{(k)}, f_{h_v}(x^{(k)})))$, or equivalently, $G_f$ is totally visible from the finite point set $\{(x^{(k)}, f_{h_v}(x^{(k)})) : x^{(k)} \in P^{(N)}\}$.*

*Proof* Since $\partial\Omega$ is $C_1$ and compact; and $f$ restricted to $\partial\Omega$ is a $C_1$-function, it follows from Lemma 2.1 that we can find a finite point set $P_1 = \{x^{(k)} \in \partial\Omega, k = 1, \ldots, M\}$ such that $\bigcup_{k=1}^{M} \mathcal{B}(x^{(k)}; \delta_{\min})$ forms a boundary layer $L_B$ about $\partial\Omega$, where $\delta_{\min}$ is the minimum radius of the open balls $\mathcal{B}(x; \delta)$ (having properties specified in Lemma 3.4) over all $x \in \partial\Omega$, and

$$L_B = \bigcup_{k=1}^{M} (\mathcal{B}(x^{(k)}; \delta_{\min}) \cap \Omega). \tag{3.18}$$

Now, consider the compact set $\Omega - L_B$. Although we know from the Heine-Borel covering theorem that there exists a finite covering of $\Omega - L_B$, but here we need to construct a covering of $\Omega - L_B$ from the projected visible sets corresponding to a finite set of points in $\Omega - L_B$. Following the foregoing approach, we can find a finite point set $P_2 = \{\tilde{x}^{(k)} \in \Omega - L_B, k = 1, \ldots, \tilde{M}\}$ with sufficiently large $\tilde{M}$, such that

$$\Omega - L_B \subseteq \bigcup_{k=1}^{\tilde{M}} \mathcal{B}(\tilde{x}^{(k)}; \delta_k) \tag{3.19}$$

with $\delta_k > 0, k = 1, \ldots, \tilde{M}$. Thus, $P_1 \cup P_2$ is a finite point-set having the desired properties. □

The significance of the foregoing result is that there exists a *finite* set of points in $\mathcal{P} = G_{f_{h_v}}$ from which the object surface $G_f$ is totally visible.

**Problem 3.4 Minimal Observation-Point Set Problem.** Given a compact connected object $\mathcal{O}$ and an observation platform $\mathcal{P}$, find a point set $\mathcal{P}^{(N)} = \{z^{(i)} \in \mathcal{P}, i = 1, \ldots, N\}$ with the smallest cardinal number $N$ such that $\mathcal{O}$ is totally visible from $\mathcal{P}^{(N)}$, or $\bigcup_{i=1}^{N} \mathcal{V}(z^{(i)}) = \partial \mathcal{O}$.

This is basically a set-covering problem [4–9]. Its complexity depends on the shape of the object under observation and the observation platform. The solution of this problem can be accomplished by decomposing it into two sub-problems:

**Problem 3.4a** Compute $\mathcal{V}(z)$ for any $\in \mathcal{P}$.

**Problem 3.4b** Find the smallest finite set $\mathcal{P}^{(N)}$ of observation points such that $\bigcup_{z^{(i)} \in \mathcal{P}^{(N)}} \mathcal{V}(z^{(i)}) = \partial \mathcal{O}$.

Problem 3.4a is a problem in computational geometry, while Problem 3.4b is a minimum set-covering problem [4, 5](see Appendix A for a brief description). If we approximate $\mathcal{P}$ by a discretized surface (e.g. a polyhedral surface) characterized by a finite point set, then Problem 3.4b becomes an optimal combinatorial set-covering problem. Evidently, the complexity of Problem 3.4 depends to a large extent on the connectedness of the object under observation.

### 3.2.1 Simply-Connected Objects

We shall examine a few simple cases for which explicit solutions to Problem 3.4 can be obtained.

   (i) **Closed Spherical Ball:** Consider a closed spherical ball $\bar{\mathcal{B}}(0; r_o)$ centered at the origin of $\mathbb{R}^3$ with radius $r_o$ and boundary sphere $\mathcal{S}(0; r_o)$.

**Proposition 3.2** *Let the object under observation be the closed spherical ball* $\bar{\mathcal{B}}(0; r_o) \subset \mathbb{R}^3$ *with observation platform being the sphere* $\mathcal{S}(0; r_o + h)$ *with radius* $r_o + h, 0 < h < \infty$, *and concentric with* $\mathcal{S}(0; r_o)$. *Then the minimum number of distinct observation points on* $\mathcal{S}(0; r_o + h)$ *for total visibility is given by* $I_s(\pi / \cos^{-1}(r_o/(r_o + h)))$, *where* $I_s(a)$ *denotes the smallest integer* $A \geq 0$ *such that* $a \leq A$.

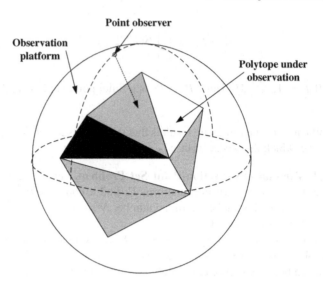

**Fig. 3.8**  Observation of a polytope

*Proof*  For the observation point $o_1 \stackrel{\text{def}}{=} (\bar{z}_1, \bar{z}_2, \bar{z}_3) = (0, r_o + h, 0) \in \mathcal{S}(0; r_o + h)$, its visible set is a spherical cap described by

$$\mathcal{V}(o_1) = \{(x_1, x_2, x_3) \in \mathbb{R}^3 : x_1^2 + x_2^2 + x_3^2 = r_o^2; x_1^2 + x_3^2 \le r_o^2 \cos^2 \alpha; r_o \sin \alpha \le x_2 \le r_o\},$$
(3.20)

where $\alpha = \sin^{-1}(r_o/(r_o + h))$ is the half aperture angle of the observation cone tangent to $\bar{B}(0; r_o)$ with vertex at $o_1$. The visible set of any observation point $z \in \mathcal{S}(0; r_o + h)$ can be obtained by an appropriate rotation of $\mathcal{V}(o_1)$ about the origin. Evidently, the minimum number of observation points on $\mathcal{S}(0; r_o + h)$ to achieve total visibility of $\mathcal{S}(0; r_o)$ is the smallest positive integer closest to $\pi/(\pi/2 - \alpha)$ or $\mathsf{I}_s(\pi/\cos^{-1}(r_o/(r_o + h)))$. Moreover, the location of the observation-point set in $\mathcal{S}(0; r_o + h)$ for total visibility is non-unique.  $\square$

*Remark 3.5*  An explicit expression for the visible set of any observation point exterior to a prolate spheroid or ellipsoid can be derived from (3.20) by introducing a nonsingular linear transformation defined by appropriate coordinate scaling, since the tangency between a line and the body surface is preserved under any nonsingular linear transformation.

**(ii) Polytopes in $\mathbb{R}^3$**: Let $\mathcal{O}$ be a $N$-faced polytope enclosed by a spherical observation platform $\mathcal{P}$ with radius $r_o$ as illustrated by Fig. 3.8.

In the special case where any point lying on or in the interior of a polyhedron is an admissible observation point, Problem 3.4 corresponds to finding the small number of admissible point-observers such that every face of the polyhedron is totally visible by at least one of the observers. This problem is akin to the "Interior Art Gallery Problem" first posed by Klee [10] (see Appendix B for a brief description). In the trivial case where the polyhedron is convex, only one point-observer is needed for

**Fig. 3.9** A star-shaped polyhedron with an interior point-
observer

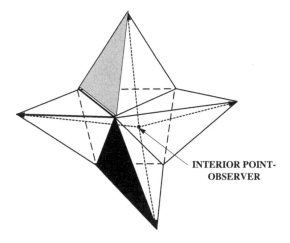

INTERIOR POINT-
OBSERVER

total visibility of the polyhedron. Note that total visibility of the polyhedron may
be attainable by only one point-observer even the polyhedron is nonconvex (e.g. a
star-shaped polyhedron as shown in Fig. 3.9).

### 3.2.2  Non-simply Connected Objects

For a 3D non-simply connected object, the determination of minimum number of
point-observers for total visibility is generally a difficult problem. We shall examine
a few cases where explicit solutions to Problem 3.4 are obtainable.

  (i) **Toroidal Objects:** First, consider the case where the solid object $\mathcal{O}$ whose
surface $\partial\mathcal{O}$ under observation is a 3-D torus described by:

$$\partial\mathcal{O} = \{(x_1, x_2, x_3) \in \mathbb{R}^3 : x_3^2 = r^2 - (R - \sqrt{x_1^2 + x_2^2})^2\}, \qquad (3.21)$$

where $r$ is the radius of the circular torus tube, and $R$ is the distance from the
torus center to the center of the tube satisfying $R > r$. The observation points are
restricted to the exterior of $\mathcal{O}$ and its boundary surface $\partial\mathcal{O}$. Moreover, we require
that the distances of the observation points $z^{(i)}$ from the torus center are $> R + r$
and $\leq \Delta$, a specified distance $> R + r$ (See Fig. 3.10). This case is relevant to the
problem of sensor placement for observing a toroidal plasma such as that in the
tokamak machine. The solution to Problem 3.4 can be constructed by making use of
the geometric symmetry of the torus with respect to the $x_3$-axis. To obtain the largest
visible sets for each observation point, two of the point observers $z^{(1)}$ and $z^{(2)}$ should
be at the maximum allowable distance $\Delta$ from the torus center on the $x_3$-axis. The
visible sets of these observation points are given by

$$\mathcal{V}(z^{(1)}) = \{(x_1, x_2, x_3) \in \mathbb{R}^3 : x_3 = \sqrt{r^2 - \left(R - \sqrt{x_1^2 + x_2^2}\right)^2},$$

$$\text{if } (R - r\cos(\theta_4))^2 < x_1^2 + x_2^2 \leq (R + r\cos(\theta_3))^2;$$

$$x_3^2 = r^2 - (R^2 - \sqrt{x_1^2 + x_2^2})^2, \text{ if } r^2 \leq x_1^2 + x_2^2 \leq (R - r\cos(\theta_4))^2\}, \qquad (3.22)$$

$$\mathcal{V}(z^{(2)}) = \{(x_1, x_2, x_3) \in \mathbb{R}^3 : x_3 = -\sqrt{r^2 - \left(R - \sqrt{x_1^2 + x_2^2}\right)^2}$$

$$\text{if } (R - r\cos(\theta_4))^2 < x_1^2 + x_2^2 \leq (R + r\cos(\theta_3))^2;$$

$$x_3^2 = r^2 - \left(R^2 - \sqrt{x_1^2 + x_2^2}\right)^2, \text{ if } r^2 \leq x_1^2 + x_2^2 \leq (R - r\cos(\theta_4))^2\}, \qquad (3.23)$$

where

$$\theta_1 = \tan^{-1}(\Delta/R), \quad \theta_2 = \cos^{-1}(r/\sqrt{\Delta^2 + R^2}), \quad \theta_3 = \pi - \theta_1 - \theta_2, \quad \theta_4 = \theta_2 - \theta_1. \qquad (3.24)$$

The invisible set of these points corresponds to $\partial\mathcal{O} - (\mathcal{V}(z^{(1)}) \cup \mathcal{V}(z^{(2)}))$ which is a circular band given by

$$B_d = \{(x_1, x_2, x_3) \in \mathbb{R}^3 : x_3^2 = r^2 - (R - \sqrt{x_1^2 + x_2^2})^2,$$

$$\text{if } (R - r\cos(\theta_3))^2 \leq x_1^2 + x_2^2 \leq (R + r)^2\}. \qquad (3.25)$$

The remaining problem is determine the minimum number of point-observers at distance $\Delta$ from the torus center to attain total visibility of $B_d$. This problem corresponds to finding a $N$-polygon whose vertices lie on the circle $\{(x_1, x_2, x_3) \in \mathbb{R}^3 : x_1^2 + x_2^2 = \Delta^2, x_3 = 0\}$ with the smallest $N$, that circumscribes the circle with radius $R + r$. The minimum number of point-observers for total visibility of $\partial\mathcal{O}$ is $2 + N$. Figure 3.10 shows the location of the point-observers for total visibility of $\partial\mathcal{O}$ for a special case. In this case, the circle $\{(x_1, x_2, x_3) \in \mathbb{R}^3 : x_1^2 + x_2^2 = (R + r)^2, x_3 = 0\}$ can be circumscribed by a square whose corners correspond to the point-observers at a distance $\Delta$ from the torus center. Thus, the minimum number of point-observers for total visibility of $\partial\mathcal{O}$ is six.

Next, we consider a variation of the foregoing case in which the object under observation is a toroidal cavity whose wall is described by (3.20). It is desirable to observe the cavity wall by means of point-observers located on the wall and in the interior of the cavity. We may classify this case as an "Interior Observation-Point Set Problem", and the former case as an "Exterior Observation-Point Set Problem".

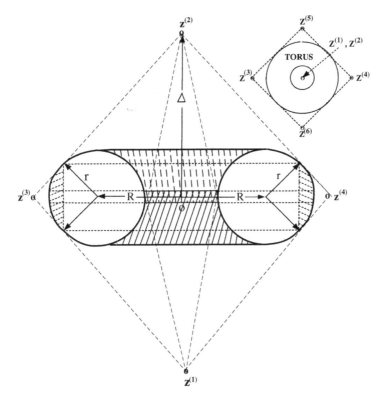

**Fig. 3.10** Minimal observation-point set for total visibility of a 3D-torus

To visualize the solution of Problem 3.4 for this case, consider first a finite-length straight cylindrical cavity whose end cross-sections are circular disks. Since the cylindrical cavity is a convex set, the line segment connecting any pair of points in the cavity lies in the cavity. Thus, the cavity wall is totally visible from any point in the interior or on the wall of the cavity. Now, suppose that the cavity is bent such that the curvatures of the cavity wall correspond to those of a torus. It is evident that there does not exist a point on the wall or in the interior of the cavity from which the cavity wall is totally visible. Moreover, an observation point $z$ having the largest visible set is one on the cavity wall that is farthermost from the origin or the torus center. Figure 3.11 shows the projection of the cavity onto the $(x_1, x_2)$-plane. The portion of the cavity surface corresponding to the projected region $\bigcup_{i=0}^{4} A_i$ is totally visible from the point-observer $z^{(1)}$ at $(x_1, x_2, x_3) = (-(R + r), 0, 0)$. Similarly, the portion of the cavity surface that is totally visible from the point-observer $z^{(2)}$ at $(x_1, x_2, x_3) = ((R + r), 0, 0)$ is indicated by the projected region $\bigcup_{i=0}^{4} B_i$. The projected regions corresponding to portions of the cavity surface that are partially visible from point-observer $z^{(1)}$ (resp. $z^{(2)}$) consist of $C_1 \cup C_2 \cup B_2 \cup B_3$ (resp. $C_1 \cup C_2 \cup A_2 \cup A_3$). The line segments $L_{Ai}$ and $L_{Bi}, i = 1, 2$, correspond to the projections of the limiting cavity cross-sections that are totally visible from

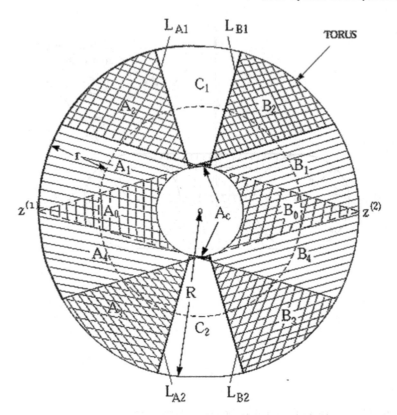

**Fig. 3.11**  Projections of visible sets of point observers for a toroidal cavity

$z^{(1)}$ and $z^{(2)}$ respectively. Finally, $A_0$ and $B_0$ correspond to the projections of the cavity surface that is visible only by $z^{(1)}$ or $z^{(2)}$ respectively. Note that there remains portions of the cavity surface (indicated by the thick black arc $A_c$) that are invisible from either point-observer $z^{(1)}$ or $z^{(2)}$. Thus, for $R > 2r$, total visibility of the cavity surface cannot be attained by two point-observers on the wall of the torus. For $2r < R \leq 3r$, the minimum number of point observers for total visibility of the cavity wall is three. Figure 3.12a shows for $R = 3r$, the locations of the point-observers for total visibility correspond to the vertices of the equilateral triangle in the $(x_1, x_2)$-plane circumscribing the circle with radius $R - r$. It can be readily verified that for $3r < R \leq (\sqrt{2}+1)/(\sqrt{2}-1)$, a minimum of four point-observers are necessary for total visibility of the cavity wall (see Fig. 3.12b). Evidently, the minimal point-observer sets for total visibility of the cavity wall are non-unique.

We may consider the minimal observation-point set problem for observing more complex toroidal objects such as multiple-hole torus from a spherical platform enclosing the objects.

**(ii) Helical Objects:** Helical objects or structures are encountered in molecular biology. A DNA molecule has the form of two intertwined helices. Also, many

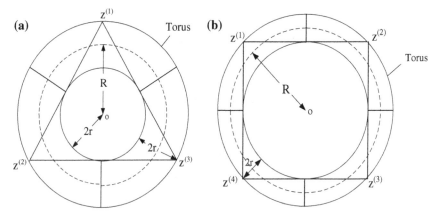

**Fig. 3.12** Projections of visible sets of point observers for a toroidal cavity; **a** $N = 3$, **b** $N = 4$

proteins have helical substructures. It is of interest to determine their structures from data acquired from a finite number of point observers. Consider a helical coil with finite length L in the world space $\mathcal{W} = \mathbb{R}^3$ with rectangular coordinates $(x_1, x_2, x_3)$. Assume that the helical coil center is a circular helix, symmetric about the $x_3$-axis with radius $r_c$ and constant pitch $2\pi b$. Moreover, the coil is formed by a solid cylinder with circular cross-section having radius $r_o$ (see Fig. 3.13). First, let the point-observer $z$ be at the origin. From the fact that the inner wall of a straight cylinder is totally visible from any point on the cylinder axis, it follows that the helical curve on the cylinder with radius $r_c - r_o$ described parametrically by

$$x_1(\lambda) = (r_c - r_o)\cos(\lambda), \quad x_2(\lambda) = (r_c - r_o)\sin(\lambda), \quad x_3(\lambda) = b\lambda, \quad 0 \leq \lambda \leq b/L \tag{3.26}$$

is totally visible from an observation-point $z$ at the origin. This also implies that the helical curve with radius $r_c$ (i.e. set $r_o = 0$) is totally visible from an observation point $z$ at the origin. For a helical coil with $r_o > 0$, the coil surface is partially visible from $z$. The visible surface area diminishes as the observed point moves away from the origin. It is evident that point observers located on the $x_3$-axis and outside the helix are necessary to achieve total visibility.

(iii) **Objects Formed by Chains of Compact Convex Sets:** Let $\mathcal{O} = \{\mathcal{O}_i, i = 1, \ldots, N\}$ be a given sequence of compact convex sets in the world space $\mathcal{W} = \mathbb{R}^n$, $n \in \{2, 3\}$. $\mathcal{O}$ is said to be an *open chain*, if for each $i \in \{1, \ldots, N-1\}$, $\mathcal{O}_{i+1} \cap \mathcal{O}_i \neq \phi$ and $\mathcal{O}_j \cap \mathcal{O}_i = \phi$ for all $j = i + 2, \ldots, N - 1$. $\mathcal{O}$ is said to be a *closed chain*, if in addition to the conditions for an open chain, $\mathcal{O}_N \cap \mathcal{O}_1 \neq \phi$. The object under observation $\mathcal{O}$ corresponds to $\bigcup_{i=1}^{N} \mathcal{O}_i$. The observations are made from points in $\mathcal{O}$. For such chains, the solutions to Problem 3.3 follow trivially from the convexity of $\mathcal{O}_i$:

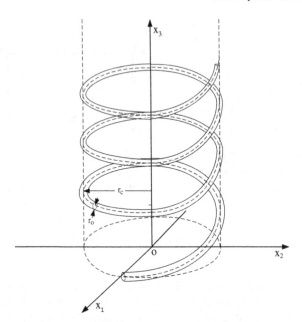

**Fig. 3.13** Helical structure in $\mathbb{R}^3$

**Proposition 3.3** *Let $Q_i = \mathcal{O}_{i+1} \cap \mathcal{O}_i$, $i = 1, \ldots, N - 1$. If $S$ is an open (resp. a closed) chain, then $\mathcal{O}$ is totally visible from a minimal set of observation points $\{z^{(1)}, \ldots, z^{(N-1)}\}$ (resp. $\{z^{(1)}, \ldots, z^{(N-2)}\}$), where $z^{(i)}$ is any point in $Q_i$.*

The foregoing result can be readily generalized to the case involving the observation of a sequence of connected compact convex sets (not necessarily a chain).

The following are two specific examples to illustrate the foregoing cases.

*Example 3.3* Let the world space $\mathcal{W} = \mathbb{R}^2$. $S_i$ is a circular disk with radius $r_i$, and with its center on a circle with radius $R > r_i$ for all $i = 1, \ldots, 5$ (see Fig. 3.14a). $\mathcal{O}$ is a closed chain corresponding to $\{S_i, i = 1, \ldots, 5\}$. Then, the minimum number ($n_{\min}$) of observation points for total visibility (indicated by small circles) is four. Evidently, similar results can be obtained for the case where $S_i$'s are replaced by closed spherical balls in $\mathbb{R}^3$. Also, the centers for the spherical balls lie on a simple closed curve.

*Example 3.4* $\mathcal{O}$ is an open chain formed by the interconnection of two closed spherical balls by a solid cylinder in $\mathbb{R}^3$ as shown in Fig. 3.14b. The minimum number ($n_{\min}$) of observation points for total visibility of $\partial\mathcal{O}$ is two.

**(iv) Objects Formed by a Finite Number of Connected Compact Convex Sets:**
Let $\{\mathcal{O}_i, i = 1, \ldots, N\}$ be a given family of connected compact convex sets in the world space $\mathcal{W} = \mathbb{R}^n$, $n \in \{2, 3\}$, and $\mathcal{O} = \bigcup_{i=1}^{N} \mathcal{O}_i$. For a given $\mathcal{O}_i$, there exists at least one $\mathcal{O}_j$, $j \in \{1, \ldots, N\}/\{i\}$ such that $\mathcal{O}_i \cap \mathcal{O}_j \neq \phi$.

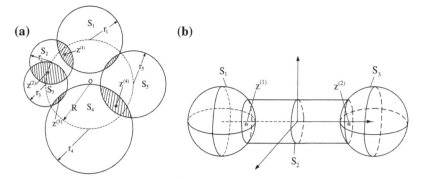

**Fig. 3.14** Chains of convex sets: **a** closed-chain of circular disks in $\mathbb{R}^2$; **b** open-chain formed by the interconnection of two closed spherical balls by a solid cylinder in $\mathbb{R}^3$

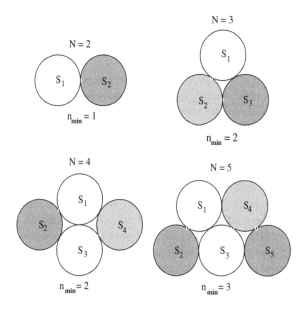

**Fig. 3.15** Close-packed closed circular disks in $\mathbb{R}^2$

*Example 3.5* Let $\mathcal{O}$ be $N$ close-packed identical closed circular disks $S_i$ in $\mathbb{R}^2$ as shown in Fig. 3.15. Obviously, the observation points with maximum visibility correspond to the points of contact between two disks or common boundary points of two disks. Thus, the minimal set of contact points that attains total visibility is a solution to Problem 3.4. These observation points are indicated by white circles. The locations of these observation points for total visibility of $\mathcal{O}$ are non-unique. Depending on the location of the disks, each $S_i$ can have at most six contact points. It can be verified that the minimum number ($n_{\min}$) of observation points for total visibility of $\mathcal{O}$ is given by

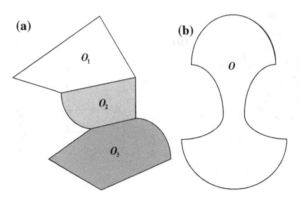

**Fig. 3.16  a** Observed object expressible as the union of three connected compact convex sets; **b** Observed object not expressible as the union of connected compact convex sets

$$n_{\min} = \begin{cases} N/2 & \text{for } N \text{ even}; \\ (N-1)/2 + 1 & \text{for } N \text{ odd.} \end{cases} \tag{3.27}$$

When the closed circular disks $S_i$'s in $\mathbb{R}^2$ are replaced by close-packed identical closed spherical balls in $\mathbb{R}^3$, each $S_i$ can have at most eight contact points. However, only two adjacent $S_i$'s are totally visible from each contact point. Consequently, the expression given by (3.26) for the minimum number of observation point for total visibility of $\mathcal{O}$ remains valid.

*Remark 3.6* Given an object $\mathcal{O}$ described by a connected compact set in $\mathbb{R}^n$, $n \in \{2, 3\}$, we may ask whether it can be expressed as the union of a finite number of connected compact convex sets $\mathcal{O}_i$ in $\mathbb{R}^n$. If so, the minimum number of interior point-observers can be readily determined; if not, we may seek an approximation of $\mathcal{O}$ by the union of a minimum number of connected compact convex sets $\mathcal{O}_i$, thereby obtaining a near-optimal solution to the Minimal Interior Point-Observer Set Problem. The accuracy of approximation of $\mathcal{O}$ may be expressed in terms of the Hausdorff distance between $\mathcal{O}$ and $\bigcup_{i=1}^{N} \mathcal{O}_i$. Figure 3.16 shows examples of the observed objects in $\mathbb{R}^2$ that are expressible or not expressible as the union of a finite number of connected compact convex sets.

*Example 3.6* Consider the object $\mathcal{O}$ in $\mathbb{R}^2$ (see Fig. 3.17) whose boundary $\partial O$ is composed of arcs $A_i$, $i = 1, \ldots, 6$. The problem is to determine the minimum number and locations of point-observers located on $\mathcal{O}^c \cup \partial \mathcal{O}$ such that the boundary of $\mathcal{O}$ is totally visible.

To solve this problem intuitively, we observe that the boundary of $\mathcal{O}$ is composed of portions of the boundary of six concatenated *fictitious objects* (circular and elliptical disks) $D_i$, $i = 1, \ldots, 6$. Evidently, the convexity of the fictitious disks implies that the boundary arcs of two neighboring fictitious disks (e.g. $A_1$ and $A_2$) are totally visible from the intersection point $p_2$ of these arcs. Thus, we can conclude that the boundary of $\mathcal{O}$ is totally visible from point-observers located at $p_2$, $p_4$ and $p_6$.

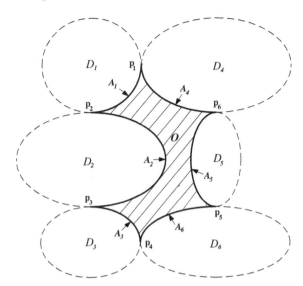

**Fig. 3.17** Object under observation in Example 3.6

## 3.3 Computational Algorithms

In practical situations, the object under observation $\mathcal{O}$ and the observation platform $\mathcal{P}$ are usually given in the form of numerical data. Approximations of $\mathcal{O}$ and $\mathcal{P}$ can be obtained by interpolation of the given numerical data. In what follows, we shall develop numerical algorithms for the approximate solution of Problems 3.1–3.3 using the numerical data directly.

Consider the simplest case discussed in Sect. 3.1, Case (i), where the observed object $\mathcal{O}$ and the observation platform $\mathcal{P}$ correspond respectively the graphs of specified real-valued $C_1$-functions $f = f(x)$ and $g = g(x)$ defined on $\Omega = [a, b]$, a compact interval of $\mathbb{R}$ satisfying $g(x) > f(x)$ for all $x \in \Omega$. Let the given numerical data be composed of the values of $f$ at uniformly spaced mesh points $x^{(i)} = a + (i - 1)\Delta x, i = 1, \ldots, N; \Delta x = (b - a)/(N - 1)$. Then, the first derivative of $f$ at $x^{(i)}$ can be approximated by the usual forward difference $Df(x^{(i)}) = (f(x^{(i+1)}) - f(x^{(i)}))/\Delta x$. For Problem 3.1, the critical height profile $h_c = h_c(x^{(i)})$ can be computed via the steps outlined in the proof of Lemma 3.1.

Next, we consider the approximate numerical solution of Problem 3.2. An essential first step is to compute the visible set of any point $z^{(i)} = (x^{(i)}, g(x^{(i)})), i = 1, \ldots, N$. This task can be accomplished by considering the points along the line segments $\mathsf{L}(z^{(i)}, y^{(j)})$ joining $z^{(i)}$ and points $y^{(j)} = (x^{(j)}, f(x^{(j)})), j = 1, \ldots, N$.

For $1 \leq j < i$, the points along the line segment $\mathsf{L}(z^{(i)}, y^{(j)})$ are given by $\{(x^{(k)}, w(x^{(k)})), j \leq k \leq i\}$, where

$$w(x^{(k)}) = g(x^{(i)}) + (f(x^{(j)}) - g(x^{(i)}))((k - i)/(j - i)), \quad j \leq k \leq i. \quad (3.28)$$

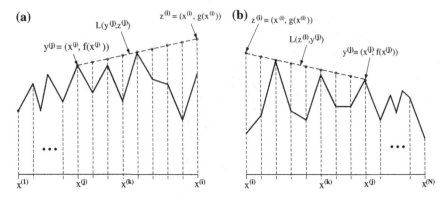

**Fig. 3.18** Determination of the points in the visible set of observation point $z^{(i)}$. The points along the line segments are indicated by *small circles*

If $w(x^{(k)}) \geq f(x^{(k)})$ for $j \leq k \leq i$, then the point $(x^{(j)}, f(x^{(j)}))$ belongs to $\mathcal{V}((x^{(i)}, g(x^{(i)})))$, the visible set of $(x^{(i)}, g(x^{(i)}))$ (see Fig.3.18a).

Similarly, for $i < j \leq N$, the points along the line segment $L(z^{(i)}, y^{(j)})$ joining $z^{(i)}$ and points $y^{(j)} = (x^{(j)}, f(x^{(j)}))$ are given by $\{(x^{(k)}, w(x^{(k)})), j \leq k \leq N\}$, where

$$w(x^{(k)}) = g(x^{(i)}) + (f(x^{(i)}) - g(x^{(j)}))((k - i)/(i - j)), \quad i \leq k \leq j. \quad (3.29)$$

If $w(x^{(k)}) \geq f(x^{(k)})$ for $i \leq k \leq j$, then the point $(x^{(j)}, f(x^{(j)}))$ belongs to $\mathcal{V}((x^{(i)}, g(x^{(i)})))$ (see Fig.3.18b).

Having computed the visible set of each point in $\mathcal{P} = G_g$, its corresponding measure as a function of $x$ can be readily determined. Thus, an approximate numerical solution to Problem 3.2 can be found simply by finding those points $x^{(i)}$ that correspond to the maximum value for the measure of the visible sets.

To obtain an approximate numerical solution to Problem 3.3, we first compute the characteristic function of $\Pi_\Omega \mathcal{V}((x^{(i)}, g(x^{(i)})))$, denoted by $\Phi(\Pi_\Omega \mathcal{V}((x^{(i)}, g(x^{(i)})))$. Now, for each point $x^{(i)} \in \Omega$, $\Phi(\Pi_\Omega \mathcal{V}((x^{(i)}, g(x^{(i)})))$ can be represented by a binary string $\mathbf{s}^{(i)}$ of length $N$. A unit string $\mathbf{s}^{(i)}$ consisting of all 1's implies that $\Omega$ is totally visible from $(x^{(i)}, g(x^{(i)}))$. If there are no such strings, we proceed by seeking pairs of distinct strings $\mathbf{s}^{(i)}$ and $\mathbf{s}^{(j)}$ such that $\mathbf{s}^{(i)} \vee \mathbf{s}^{(j)}$ is equal to the unit string, where $\vee$ denotes the logic "OR" operation between the corresponding components of $\mathbf{s}^{(i)}$ and $\mathbf{s}^{(j)}$. If there are no such string pairs, we seek triplets of distinct strings $\mathbf{s}^{(i)}, \mathbf{s}^{(j)}, \mathbf{s}^{(k)}$ such that $\mathbf{s}^{(i)} \vee \mathbf{s}^{(j)} \vee \mathbf{s}^{(k)}$ is equal to a unit string. This process is continued until a finite set of strings $\mathbf{s}^{(i)}, i = 1, \ldots, P$ such that a unit string $\bigvee_{i=1}^{P} \mathbf{s}^{(i)}$ is found. Clearly, the smallest set of such strings is an approximate numerical solution to Problem 3.3.

For the case where the given numerical data are composed of the values of $f$ and $g$ at specified mesh points $x^{(i)}$ in a two-dimensional domain $\Omega$, approximate numerical solution to Problem 3.3 can be obtained via the following steps:

(i) Approximate $G_f$ and $G_g$ by polyhedral surfaces $\hat{G}_f$ and $\hat{G}_g$ respectively.

(ii) Compute the visible sets corresponding to the vertex points in $\hat{G}_g$.

(iii) Compute the projections of the visible sets on $\Omega$, and their measures.

(iv) Determine a minimal mesh-point set such that the union of the corresponding visible sets is equal to $\Omega$.

Step (i) can be accomplished by using Delaunay triangulation to obtain approximate surfaces in the form of triangular patches (see Appendix C). Step (ii) corresponds to a problem in computational geometry involving the intersection of a flat cone (with its vertex at an observation point on $G_g$) and a triangular patch on $\hat{G}_f$ in $\mathbb{R}^3$ [11]. Step (iii) involves straightforward computation. The final step (iv) corresponds to a "Set Covering Problem" (see Appendix A) which can be reformulated as an integer programming problem. It has been shown by Cole and Sharir [12] that this problem (with $\hat{G}_g$ coinciding with the approximate observed surface $\hat{G}_f$) is NP-hard. An algorithm integrating the foregoing steps has been developed by Balmes and Wang [13]. The general idea behind this algorithm is to hop over all the observation points $z^{(i)} \in \hat{G}_g$ and try to determine whether or not a specific triangle is visible.

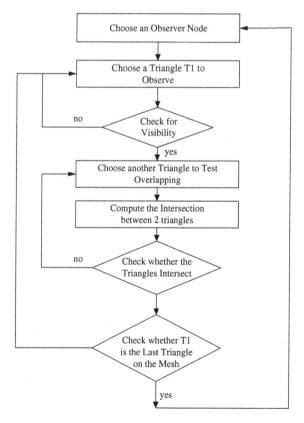

**Fig. 3.19** Flow chart of the algorithm by Balmes and Wang for computing the visible set

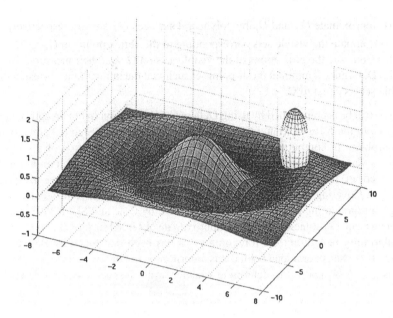

**Fig. 3.20** Simplification of triangle–triangle intersection test

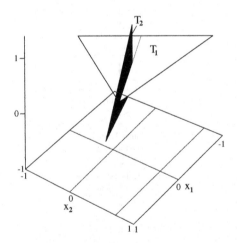

**Fig. 3.21** Sketch of two intersecting triangles

Thus, the problem is reduced to finding all the triangles that are visible from an observation point. This approach allows us to use parallel computation, since all the observation points $z^{(i)}$ are independent from each other. A flow chart of this algorithm is shown in Fig. 3.19. Since the computation involves all possible observation points and triangular patches, the time complexity of the algorithm is $O(np^2)$, where $n$ is the number of observer points vertices of the triangular patches in $\hat{G}_g$), and $p$ is the number of triangular patches. To reduce the time complexity of the algorithm, we only consider possible triangle-triangle intersections among all the triangles that are inside the sphere centered at the observation point $z^{(i)} \in \hat{G}_g$ with radius $r_o$ (see Fig 3.20). The idea here is to reduce the number of triangle-triangle intersection tests for removing the non-intersecting triangles. By considering the distance $\rho(z^{(i)}, T_1)$ between the observation point $z^{(i)}$ and a triangle $T_1$, all the triangles $T_k$ that could overlap with $T_1$ must have their distances $\rho(z^{(i)}, T_k) < \rho(z^{(i)}, T_1)$. The remaining triangles need not be tested because they do not intersect $T_1$. This fact allows the reduction of time complexity of the algorithm from $O(np^2)$ to $O(n\frac{p(p-1)}{4})$. In the triangle-triangle intersection test, all possible cases of intersection are considered, and a fast algorithm is developed for each case. This is accomplished by modifying an algorithm of Moller [14]. The basic steps of this algorithm are as follows:

Let $T_1$ and $T_2$ denote the two triangles whose intersection is to be determined. We observe that in order to have an intersection, $T_1$ and $T_2$ must be either on the same plane or at least two points of $T_1$ are strictly on both sides of the plane generated

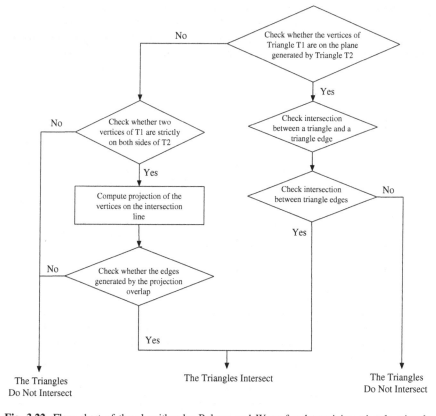

**Fig. 3.22** Flow chart of the algorithm by Balmes and Wang for determining triangle–triangle intersection

by $T_2$. A possible orientation of an intersecting triangle pair is shown in Fig. 3.21. In this configuration, we project one of the triangles to the line of intersection and make a segment overlapping test on that line. If the segments overlap, then the two triangles intersect. Otherwise, they do not intersect. A flow chart of this algorithm is shown in Fig. 3.22. The details are given in [13].

## 3.4 Numerical Examples

In what follows, we shall present a few numerical examples to illustrate the application of some of the algorithms discussed in Sect. 3.3.

*Example 3.7* **Optimal Sensor Placement in Micromachined Structures.** Consider the optimal sensor placement problem for a model of a one-dimensional micromachined solid structure whose spatial profile $G_f$ and observation platform $G_g$ are shown in Fig. 3.23, where $g = f_{h_v}$. The critical vertical-height profile $G_{h_{vc}}$ for $f$ computed by the steps outlined in the proof of Lemma 3.1 is also shown in Fig. 3.23. Since $G_{h_{vc}} \cap G_g$ is empty, $G_f$ is not totally visible from any point in $G_g$. The projection of the visible set $\mathcal{V}((x^{(i)}, g(x^{(i)})))$ on the spatial domain $\Omega = [0, 20]\ \mu$m as a function of the normalized $x^{(i)}$ (graph of the set-valued mapping $x^{(i)} \to \Pi_\Omega \mathcal{V}((x^{(i)}, g(x^{(i)})))$)

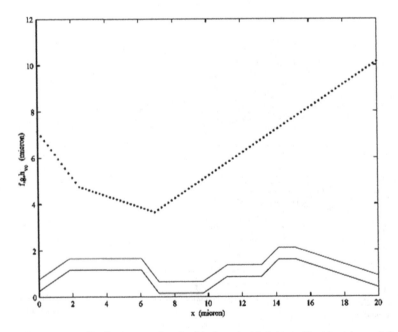

**Fig. 3.23** Spatial profile (*lower curve*) and critical vertical-height profile (*dotted curve*) for an one-dimensional micromachined structure with an elevated observation platform (*middle curve*)

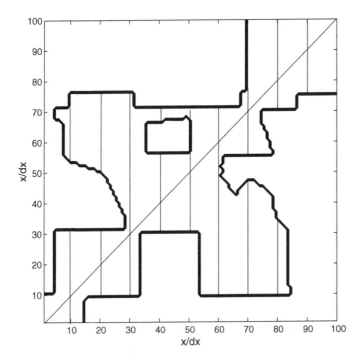

**Fig. 3.24** Projection of the visible set $\mathcal{V}((x^{(i)}, g(x^{(i)})))$ on $\Omega$ as a function of the normalized $x^{(i)}$ ($x^{(i)}/dx$, $dx = 5\,\mu$ )

on $\Omega$ into $\Omega$) is shown in Fig. 3.24, where $x^{(i)}$ is the $i$th mesh point. The corresponding measure as a function of $x^{(i)}$ is shown in Fig. 3.25. We note from Fig. 3.24 that every point in the diagonal line lies in its visible set, or every $x \in \Omega$ is a fixed point of $\Pi_I \mathcal{V}((\cdot, g(\cdot)))$ as expected. Moreover, at least three sensors are needed to cover the entire $G_f$, but their locations are non-unique. From Fig. 3.25, it is evident that the solution to the approximate optimal sensor placement problem is given by $x^* = 13.5$ $\mu$m and $\mu_1\{\Pi_\Omega \mathcal{V}((x^*, g(x^*)))\} = 16.75\,\mu$m.

Next, we consider a more complex structure formed by micromachined components having various shapes embedded in a flat bottom plane. It is required to determine the minimum number and locations of optical sensors attached to a platform above the observed surface for health monitoring and inter-structure communication. Figures 3.26 and 3.27 show the observed surface $G_f$ formed by the bottom plane and micromachined components with various geometric shapes. Here, two different observation platforms are considered. The first one corresponds to setting the sensors at a fixed distance ($10\,\mu$) above the observed surface. This case is motivated by the fact that micromachined structures are usually fabricated in layers by etching. The second observation platform corresponds to the case where the sensors lie in a plane at 5 microns above the observed surface. These two cases represent the most important ones for monitoring a micromachined structure. Evidently, the minimum

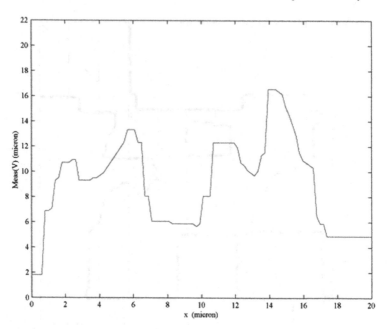

**Fig. 3.25**  Measure of the projection of visible set $\mathcal{V}((x^{(i)}, g(x^{(i)}))$ on $\Omega$ as a function of $x^{(i)}$

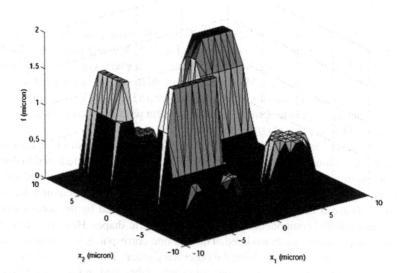

**Fig. 3.26**  Surface profile for a complex micromachined structure

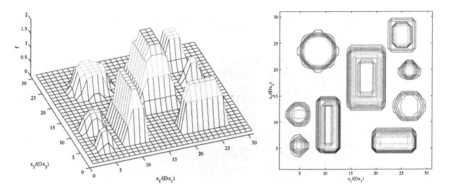

**Fig. 3.27** Spatial profile (*left figure*) and level contour (*right figure*) for the complex micromachined structure shown in Fig. 3.26; $\Delta x_1 = \Delta x_2 = 2/3\ \mu$

number and locations of the sensors for total visibility are not obvious intuitively. Figure 3.28a, b show respectively the surfaces of visible-set measures for the first and second cases respectively. In the first case, the salient features of the observed surface are also reproduced in the surface of visible-set measures, but their order and position are shifted. For the second case, there is not much to say except that the visible sets of the observation points above the higher part of the observed surface have the smallest measure.

*Example 3.8* **Optimal Placement of Optical Repeaters for Mars Surface.** In the exploration of planetary surfaces such as that of Mars using mobile robots, it is desirable to maintain communication between the robots at all times. Since the planetary surface is generally irregular, a possible option is to set up a system of optical or radio repeaters at suitable locations so that a robot is capable of communicating with any other robot. Here, we have created a simple analytical model of the Mars surface consisting of double craters which are actually found in the Mars Viking Orbiter images for "Olympic Muns" whose photograph is shown in Fig. 3.29. Here we model

**(a)**                                            **(b)**

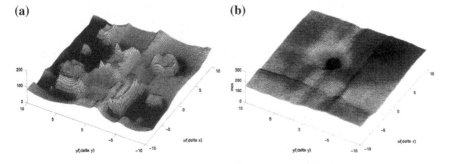

**Fig. 3.28** Surfaces of visible-set measures for the micromachined structure in Fig. 3.26 with two different forms of observation platform; $\Delta x_1 = \Delta x_2 = 2/3\ \mu$

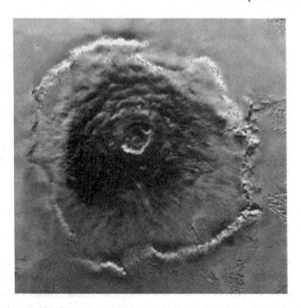

**Fig. 3.29** Photograph of Mars Olympus Muns

the double craters by the graph $G_f$ of a real-valued function $f$ described by

$$f(x_1, x_2) = f_1(x_1, x_2) + f_2(x_1, x_2), \tag{3.30}$$

where $f_i$ is the sum of two Gaussian functions:

$$f_i(x_1, x_2) = \lambda_i \left\{ \exp\left( \frac{(x_1 - x_{1ci})^2}{\sigma_{1i}^2} \right) + \exp\left( \frac{(x_2 - x_{2ci})^2}{\sigma_{2i}^2} \right) \right\}, \quad i = 1, 2, \tag{3.31}$$

in which the parameter values for the model are given by

$$\begin{bmatrix} x_{1c1} \\ x_{1c2} \end{bmatrix} = \begin{bmatrix} 1.6 \\ 1.35 \end{bmatrix}; \begin{bmatrix} x_{2c1} \\ x_{2c2} \end{bmatrix} = \begin{bmatrix} 1.6 \\ 1.35 \end{bmatrix}; \begin{bmatrix} \sigma_1 \\ \sigma_2 \end{bmatrix} = \begin{bmatrix} 0.9 \\ 0.45 \end{bmatrix}; \begin{bmatrix} \lambda_1 \\ \lambda_2 \end{bmatrix} = \begin{bmatrix} 0.75 \\ 0.45 \end{bmatrix}. \tag{3.32}$$

Figure 3.30a shows the spatial profile of the model Mars surface. Its level contours are shown in Fig. 3.30b. This model surface has features which are not found in the real Mars surface. One of them is that the craters look like waves. Moreover, the model surface does not take into account the curvature of the Mars surface. Nevertheless, the solution of the optimal repeater allocation problem using this model provides some information on the optimal locations of the repeaters that are not obvious by intuitive reasoning. Assuming that all the repeaters have fixed vertical height $h_v$, the observation platform $G_g$ corresponds to the elevated surface $G_{f_{h_v}}$. Here, the visible set of a point on the approximate surface $\hat{G}_g$ is a subset of $\hat{G}_f$. Using the

(a)                                          (b)

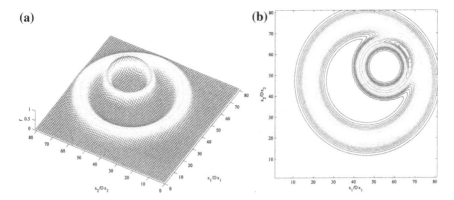

**Fig. 3.30** **a** Surface model of Mars Olympus Muns. **b** Level contours of the Mars Olympus Muns surface model; $\Delta x_1 = \Delta x_2 = 0.25\,\text{km}$

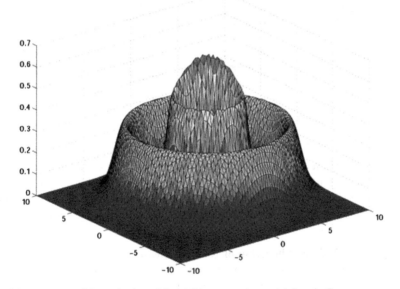

**Fig. 3.31** Measure of the projection of the visible sets on the spatial domain $\Omega$

Balmes-Wang algorithm described in Sect. 3.3, the measures of the projections of the visible sets on the spatial domain $\Omega$ are computed for all the mesh points. The results are shown in Fig. 3.31. As expected the computed surface has almost the same symmetry as that of the model Mars surface. There is apparently no reason for breaking the symmetry when one computes the elements visible from an observation point. Moreover, for this surface, we did not take into account the fact that in a corner, we have many points whose visible sets have high measures because they correspond to the flat part of the surface and not the craters. The computed results show that ten

**(a)** **(b)**

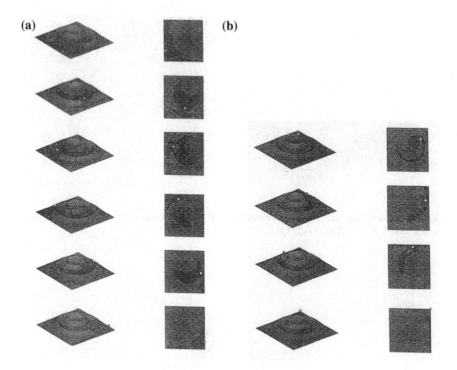

**Fig. 3.32** Alternate set of repeaters for Mars surface

repeaters are sufficient to cover the entire surface. This a sub-optimal solution. The locations of these repeaters are not unique (see Fig. 3.32). Nevertheless the solution is non-trivial. Despite of the symmetrical features of the surface, a large number of observation points or repeaters are needed for total coverage of the surface.

*Example 3.9* **Optimal Allocation of Radio Repeaters.** An important problem in the design of a wireless communication network is to determine the minimum number and locations of radio repeaters to attain total coverage of a given spatial terrain. We assume that the antennas for the repeaters have fixed heights. Here, the actual elevation data for the Los Angeles area obtained from the U.S. National Geographical Data Center are used. The approximate terrain is generated from the actual data by sampling at 5-minute intervals and applying Delaunay triangulation to obtain a suitable mesh for reconstructing the surface by triangular patches. Although the accuracy of the approximation is not very high, it is sufficient to preserve the dominant features of the surface. Figure 3.33 shows the reconstructed spatial terrain profile and its contours. It can be seen from Fig. 3.33 that the data also include the elevation of the sea. This is not so important for computing the visible sets. But in solving the optimization problem, it is necessary to remove the sea region if we wish to consider only land-based radio repeaters. Figure 3.34 shows the surface of visible-set measures for an

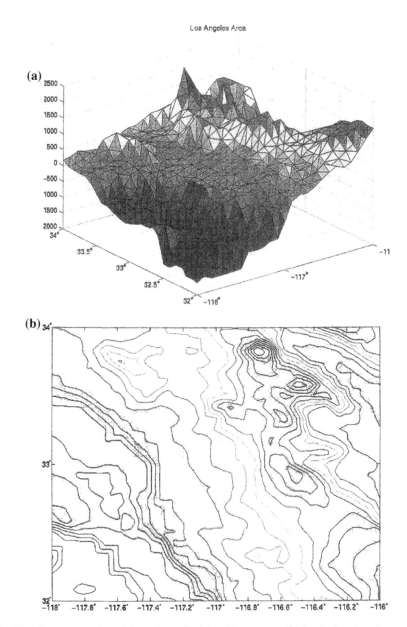

**Fig. 3.33** Reconstructed spatial terrain profile (**a**) and its contours (**b**) for the Los Angeles area

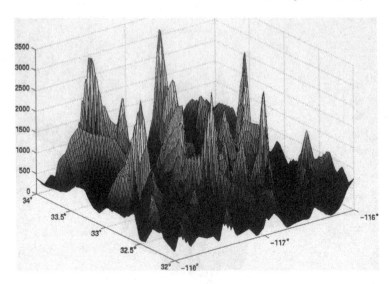

**Fig. 3.34** Measure of the visible-set measures for an observation platform with fixed vertical-height at 10 ft

observation platform with fixed vertical-height at 10 ft. computed using the Balmes-Wang algorithm. The results reveal that the lack of structure and symmetry of the surface generally leads to non-unique solutions to the optimal repeater allocation problem. Also, a minimum of 23 repeaters are needed to cover the entire Los Angeles area.

## Exercises

**Ex. 3.1.** Let the object $\mathcal{O}$ under observation be a plane curve described by the graph $G_f$ of the real-valued function $f(x) = 1 + \cos(x)$ defined on $\Omega = [-\pi, \pi] \subset \mathbb{R}$. The point-observers $z$ lie in $\text{Epi}_f$.

(a) Determine the critical vertical height $h_{vc}(x)$ for any $x \in b\Omega$.
(b) Determine the visible set $\mathcal{V}(z)$ for any observation point $z = (x, f(x) + h_v(x))$ such that $0 < h_v(x) < h_{vc}(x)$ for any $x \in \Omega$.
(c) Let the observation platform $\mathcal{P}$ be $G_{f+h_v}$, the graph of $f + h_v$, where $h_v$ is a constant vertical height $< h_{vc}(x)$ for all $x \in \Omega$. Find the observation points $z^* \in \mathcal{P}$ such that $\mu_1\{\Pi_\Omega \mathcal{V}(z^*)\} \geq \mu_1\{\Pi_\Omega \mathcal{V}(z)\}$ for all $z \in \mathcal{P}$, where $\mu\mu_1\{A\}$ denotes the Lebesgue measure of $A$. Is $z^*$ unique?
(d) Repeat parts (a)–(c) for the object $\mathcal{O}$ described by $G_{\tilde{f}}$, where $\tilde{f}(x) = 1 - \cos(x)$, $x \in \Omega$.

**Ex. 3.2.** The sketch of a rocket composed of a cylinder with circular cross-section and length $L_2$, a nose cone and a truncated conical-shaped engine as shown in Fig. 3.35. The objective is to place cameras modelled by point observers along the observation platform $\mathcal{P}$ represented by the red circle at the rocket base so that the entire rocket exterior can be observed.

**Fig. 3.35** Sketch of a rocket
for Exercise 3.2

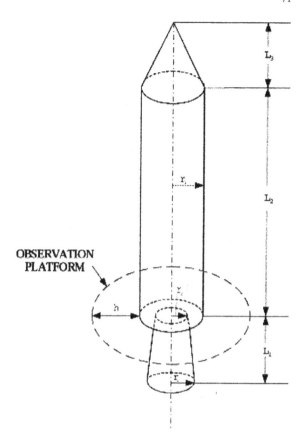

(a) Given a point observer $z$ on the observation platform $\mathcal{P}_h$ with height $h$, find its visible set.

(b) For a given point observer $z \in \mathcal{P}_h$, determine the minimal set of point observers to achieve total visibility of the rocket exterior surface.

**Ex.3.3.** Two different configurations for a space habitat are shown in Fig. 3.36. From the geometric properties of each habitat, use your intuition to determine qualitatively a minimal set of point observers located inside the habitat for total visibility of the habitat interior. Give your reasoning for arriving at the conclusion that the set is minimal.

**Ex.3.4.** Let the object $\mathcal{O}$ under observation be composed of a spherical ball and a torus with rectangular cross-section as shown in Fig. 3.37.

(a) Find a minimal set of observation points in the exterior of $\mathcal{O}$ for total visibility of $\partial\mathcal{O}$, the boundary surface of $\mathcal{O}$.

(b) Find a minimal set of observation points inside the toroidal cavity for total visibility of the cavity wall.

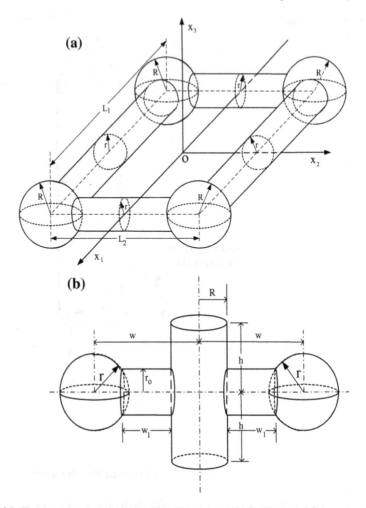

**Fig. 3.36** Sketches of space-habitat configurations (**a**) and (**b**) for Exercise 3.3

**Ex.3.5.** Figure 3.38 shows a truncated cone with aperture angle $\theta$ and length $L$, attached to a spherical ball with radius $r_o = L \tan(\theta/2)$. The observation platform $\mathcal{P}$ is a circle with radius $R > r_o$. Determine whether total visibility of $\mathcal{O}$ is attainable by a finite set of observations in $\mathcal{P}$.

**Ex.3.6.** A bar with a rectangular cross-section with $L \gg W > T$ as shown in Fig. 3.39, where $L$, $W$ and $T$ are the length, width and thickness respectively.

(a) Show that totally visibility of the bar's surface can be attained by a pair of point-observers located at finite distances from the bar. Give the location of each point-observer.

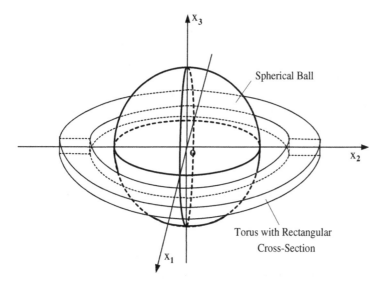

**Fig. 3.37** Sketch of the object under observation for Exercise 3.4

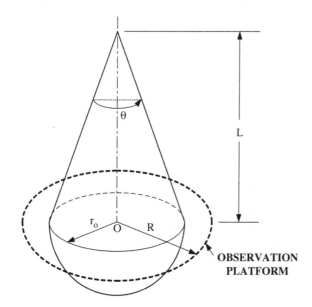

**Fig. 3.38** Sketch of the object under observation for Exercise 3.5

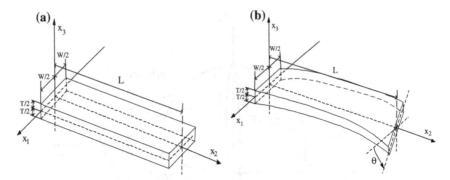

**Fig. 3.39 a** Sketch of the bar with rectangular cross-section. **b** Twisted bar with rectangular cross-section

(b) Now, we twist the bar about the $x_2$-axis according to $\theta(x_2) = \sin(x_2\pi/2L)$ as shown in Fig. 3.33b. Determine whether total visibility of the bar's surface can be attained by a pair of point-observers at finite distances from the bar. If so, give the location of each point-observer. If not, explain why not.

(c) Now, we connect the two ends of the twisted bar in part (b) to form a Mobiüs strip. Determine whether it is possible to attain total visibility of the Mobiüs bar using a pair of point-observers?

**Ex.3.7.** Consider an object $\mathcal{O}$ composed of the union of two interlaced identical toroid as shown in Fig. 3.40. Determine the minimal set of observation points in $\mathcal{O}^c$ for total visibility of $\mathcal{O}$.

**Ex.3.8.** Figure 3.41 shows a sketch of spherical storage tanks for an oil refinery. The tanks are supported by cylindrical columns with length $L$ and their axes located at the vertices of an equilateral triangle with side $a$. Determine a minimal set of observation points exterior to the structure for total visibility.

**Ex.3.9.** Consider the objects $\mathcal{O}$ in $\mathbb{R}^2$ whose boundaries are composed of line segments and circular arcs as shown in Fig. 3.42(a)–(d). Determine which ones have Property A: The object can be expressed as the union of a finite number of connected compact convex sets. For those objects having Property A, determine the minimum number of point-observer in $\bar{\mathcal{O}}$ for total visibility of the interior of $\mathcal{O}$. For each object that does not have Property A, construct an approximation of the object that has Property A.

**Ex.3.10.** Make use of the notion of *fictitious* objects introduced in Example 3.6 to determine the minimal number and locations of point-observers in $\partial\mathcal{O} \cup \mathcal{O}^c$ for total visibility of the boundary of each object $\mathcal{O}$ shown in Fig. 3.43.

**Ex.3.11.** Consider an opaque Möbius strip with zero thickness (Fig. 3.44) described by:

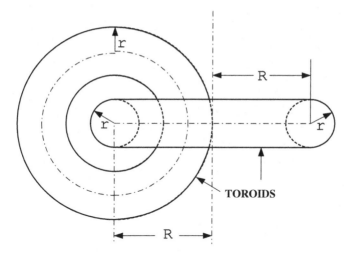

**Fig. 3.40** Sketch of two interlaced toroids for Exercise 3.5. *Note* Here the object is not a chain as defined earlier, since the intersection of the toroids is empty

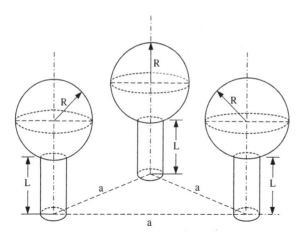

**Fig. 3.41** Sketch of storage tanks for an oil refinery

$$x_1 = (R + \alpha \cos(\beta/2)) \cos(\beta), \quad x_2 = (R + \alpha \sin(\beta/2)) \sin(\beta), \quad x_3 = \alpha \sin(\beta/2),$$
$$(3.33)$$

where $R$ is the mid-circle radius at $x_3 = 0$, and $W$ is the half-width of the strip. Determine the minimum number of point-observers located outside the strip to achieve total visibility. Note: For a zero-thickness strip, ray initiated from the point-observer can approach the surface from either side of the strip.

**Ex.3.12.** A bracelet $\mathcal{O}$ made from $N \geq 3$ identical spherical gemstones with radius $R_o$ is illustrated in Fig. 3.45. Each gemstone has a hole through its diameter so that

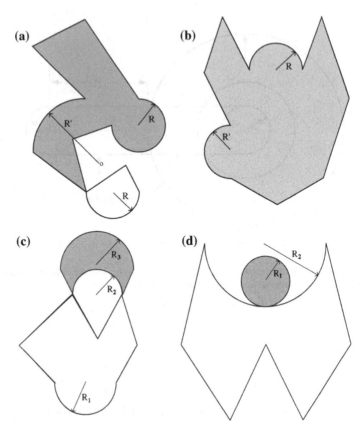

**Fig. 3.42** Compact objects in $\mathbb{R}^2$

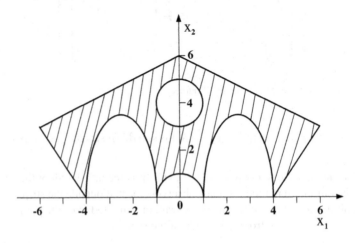

**Fig. 3.43** Objects in $\mathbb{R}^2$

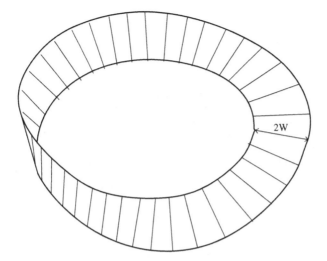

**Fig. 3.44** A zero-thickness opaque Möbius strip under observation

**Fig. 3.45** A bracelet made of $N$ identical spherical gemstones (shown for $N = 3$)

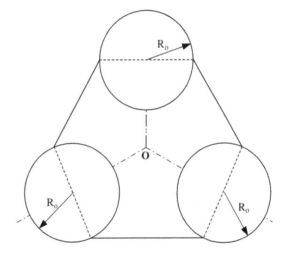

the gemstones can be linked by a string (assume zero volume for simplicity). Thus, $\mathcal{O}$ corresponds to equally-spaced spheres in $\mathbb{R}^3$ linked by a polygonal string-loop in $\mathbb{R}^2$ with diameter $\geq 3R_o$. The problem is to find a minimal finite set of observation points at finite distances from the $\mathcal{O}$ such that total visibility of the gemstones' surfaces is attained.

# References

1. P.K.C. Wang, A class of optimization problems involving set measures. Nonlinear Anal. Theor. Methods Appl. **47**, 25–36 (2001)
2. V.F. Dem'yanov and L.V. Vasil'ev, Nondifferentiable Optimization, Optimization Software, New York, 1985 (translated from Russian by T. Sasagawa)
3. F.H. Clarke, Methods of Dynamic and Nonsmooth Optimization, Regional Conference Series in Applied Math. Soc. for Industrial and Applied Math. Philadelphia, 1989
4. M. Demange, V. Paschos, *Approximation Algorithms for Minimum Set Covering Problem: A Survey* (Université de Paris I, Patheon-Sorbonne, Paris, 1994)
5. E. Balas, A. Ho, Set covering optimization: a computational study. Math. Program. Study **12**, 37–60 (1980)
6. U. Aickelin, An indirect genetic algorithm for set covering problem. J. Oper. Res. Soc. **46**, 1260–1268 (1995)
7. M. Afif et al., A new efficient heuristic for the minimum set covering problem. J. Oper. Res. Soc. **53**, 1118–1126 (2002)
8. J.E. Beasley, An algorithm for set covering problem. Eur. J. Oper. Res. **31–1**, 85–93 (1987)
9. P. Galinier, A. Hertz, Solution techniques for the large set covering problem. Discrete Appl. Math. **155–3**, 312–326 (2007)
10. J. O'Rourke, *Art Gallery Theorems and Algorithms* (Oxford University Press, Oxford, 1987)
11. M. De Berg, *Ray Shooting, Depth Orders and Hidden Surface Removal, Lecture Notes in Computer Science* (Springer, New York, 1993)
12. R. Cole, M. Sharir, Visibility problem for polyhedral terrains. J. Symbolic Comput. **7**, 11–30 (1989)
13. C.S. Balmes, P.K.C. Wang, Numerical Algorithms for Optimal Visibility Problems, UCLA Engr. Rpt. ENG 00-214, (2000)
14. T. Moller, A fast triangle-triangle intersection test. J. Graph. Tools **2**(2), 25–30 (1997)

# Chapter 4
# Visibility-Based Optimal Path Planning

In Chap. 3, we considered various optimization problems for one or more point-observers which are stationary with respect to the object $\mathcal{O}$ under observation in the world space $\mathcal{W}$. For a single stationary point-observer, total visibility of $\mathcal{O}$ may not be attainable. A possible way to achieve total visibility of $\mathcal{O}$ is to allow the point-observer to move along some path $\Gamma_{\mathcal{P}}$ on a given observation platform $\mathcal{P}$ starting and ending at specified points $z_o, z_f \in \mathcal{P}$ respectively as illustrated in Fig. 4.1. Here, we assume that the movement of the point-observer does not involve time and dynamical effects. In effect, the single point-observer is replaced by a spatially distributed observer consisting of an uncountably infinite number of observation points along $\Gamma_{\mathcal{P}}$. A general path optimization problem is to choose an admissible path $\Gamma_{\mathcal{P}}$ in some specified class to maximize the visibility of $\mathcal{O}$ in some sense.

To formulate path optimization problems, it is necessary to specify the set of all admissible paths from which selections can be made.

**Definition 4.1** Given an observation platform $\mathcal{P} \subset \mathcal{W}$, a path $\Gamma_{\mathcal{P}}$ is said to be *admissible*, if it is a directed Jordan (simple) arc connecting the starting-point $z_o$ and end-point $z_f$. Moreover, $\Gamma_{\mathcal{P}} \subset \mathcal{P}$. The *set of all admissible paths in $\mathcal{P}$* is denoted by $\mathcal{A}_{\mathcal{P}}$.

In what follows, we shall begin our discussion with optimal path planning problems for observing two-dimensional objects in the world space $\mathcal{W} = \mathbb{R}^3$ such as surfaces or terrains. Then, similar problems for observing three-dimensional objects in $\mathcal{W} = \mathbb{R}^3$ such as solid bodies will be considered.

## 4.1 Observation of Two-Dimensional Objects

First, consider the case where the object under observation $\mathcal{O}$ and the observation platform $\mathcal{P}$ are described respectively by $G_f$ and $G_g$ (the graphs of given real-valued $C_1$-functions $f = f(x)$ and $g = g(x)$ defined on a given compact set $\Omega \subset \mathbb{R}^2$

© Springer International Publishing Switzerland 2015
P.K.-C. Wang, *Visibility-based Optimal Path and Motion Planning*,
Studies in Computational Intelligence 568, DOI 10.1007/978-3-319-09779-4_4

**Fig. 4.1** Admissible path for
a mobile point-observer

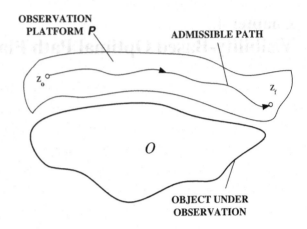

OBSERVATION
PLATFORM $P$

ADMISSIBLE PATH

$z_0$

$z_f$

$O$

OBJECT UNDER
OBSERVATION

satisfying $g(x) > f(x)$ for all $x \in \Omega$). For this case, along any admissible path $\Gamma_{\mathcal{P}} \subset \mathcal{P}$, the corresponding arc $\mathcal{C}(\Gamma_{\mathcal{P}}) = \{(x, f(x)) : (x, g(x)) \in \Gamma_{\mathcal{P}}\} \subset G_f$ is a Jordan arc, and so is $\Pi_{\Omega}\mathcal{C}(\Gamma_{\mathcal{P}})$, the projection of $\mathcal{C}(\Gamma_{\mathcal{P}})$ on $\Omega$.

**Problem 4.1** *Single Mobile Point-Observer Shortest Path Problem.* Given an observation platform $\mathcal{P} = G_g$ and two distinct points $z_o = (x_o, g(x_o))$, $z_f = (x_f, g(x_f)) \in \mathcal{P}$, find the shortest admissible path $\Gamma_{\mathcal{P}}^* \in \mathcal{A}_{\mathcal{P}}$ starting at $z_o$ and ending at $z_f$ such that

$$\bigcup_{z \in \Gamma_{\mathcal{P}}^*} \mathcal{V}(z) = G_f. \tag{4.1}$$

In many situations, instead of considering admissible paths $\Gamma_{\mathcal{P}}$ in $\mathcal{A}_{\mathcal{P}}$, it is more convenient to consider admissible paths $\Gamma \in \mathcal{A}_{\Omega}$ (the set of all admissible paths $\Gamma$ in $\Omega$). Thus, we have the following modified version of Problem 4.1.

**Problem 4.1'** Given an observation platform $\mathcal{P} = G_g$ and $z_o = (x_o, g(x_o))$, $z_f = (x_f, g(x_f)) \in \mathcal{P}$ such that $x_o \neq x_f$, find the shortest admissible path $\Gamma^* \in \mathcal{A}_{\Omega}$ starting at $x_o$ and ending at $x_f$ such that

$$\bigcup_{x \in \Gamma^*} \Pi_{\Omega}\mathcal{V}((x, g(x))) = \Omega. \tag{4.2}$$

*Remark 4.1* Evidently, the shortest path $\Gamma^*$ in $\Omega$ generally does not imply that the corresponding path $\Gamma_{\mathcal{P}}$ in $\mathcal{P}$ has the shortest length, hence Problems 4.1 and 4.1' are generally not equivalent. Nevertheless, it is still useful to consider Problem 4.1', since its solution provides insight into the solution of corresponding Problem 4.1. Next, we observe that for $\mathcal{P} = G_{f+h_v}$ (the constant vertical-height observation platform), once a solution to Problem 3.5 (Minimal Observation-Point Set Problem) is obtained, any admissible path starting at $z_o$, passing through all the points in the observation-point set $\mathcal{P}^{(N)}$, and ending at $z_f$, is a candidate to the solution of Problem 4.1.

In planetary surface exploration, it is important to avoid paths with steep slopes. This requirement can be satisfied by including the following gradient constraint in Problem 4.1:

$$\|\nabla f(x)\| \leq f'_{\max} \quad \text{for all } x \in \Gamma^*, \tag{4.3}$$

where $f'_{\max}$ is a specified positive number. This modified problem will be referred to hereafter as Problem 4.1".

**Problem 4.2** *Maximum Visibility Problem.* Let $x_o$ and $x_f$ be specified distinct points in $\Omega$. Let $\tilde{\mathcal{A}}_\Omega$ denote the set of all admissible paths $\Gamma$ from $x_o$ to $x_f$ satisfying the arc length constraint:

$$0 < l_{\min} \leq \Lambda(\Gamma) \leq l_{\max} < \infty, \tag{4.4}$$

where $l_{\max}$ and $l_{\min}$ are specified positive numbers, and $\Lambda(\Gamma)$ denotes the arc length of $\Gamma$. Find a $\Gamma^* \in \tilde{\mathcal{A}}_\Omega$ such that the path visibility functional defined by the following line integral from $x_o$ to $x_f$ along arc $\Gamma$

$$J_1(\Gamma) = \int\limits_0^{\Lambda(\Gamma)} \mu_2\{\Pi_\Omega \mathcal{V}((x, f_{h_v}(x))|_{x \in \Gamma})\} dl \tag{4.5}$$

satisfies $J_1(\Gamma^*) \geq J_1(\Gamma)$ for all $\Gamma \in \tilde{\mathcal{A}}_\Omega$, where $dl$ is the elemental arc length, $f_{h_v}(x) \overset{\text{def}}{=} f(x) + h_v$, and $\mu_2\{S\}$ denotes the Lebesgue measure or surface area of $S \subset \mathbb{R}^2$. In the case where the line segment joining $x_o$ and $x_f$ lies in $\Omega$, we set $l_{\min} = \|x_f - x_o\|$.

Another physically meaningful path visibility functional is given by

$$J_2(\Gamma) = \mu_2\{\bigcup_{x \in \Gamma} \Pi_\Omega \mathcal{V}((x, f_{h_v}(x)))\}. \tag{4.6}$$

Problem 4.2 with $J_1$ replaced by $J_2$ corresponds to selecting an admissible path $\Gamma$ such that the measure of the union of the projected visibility sets on $\Omega$ for all the points along $\Gamma$ is maximized. This modified problem will be referred hereafter as Problem 4.2".

*Remark 4.2* Both Problems 4.1' and 4.2 involve admissible paths in $\Omega$. If we consider paths corresponding to Jordan arcs in $G_g$ connecting points $z_o$ and $z_f$ (e.g. a mobile robot observer's paths are confined to a planetary surface described by $G_{f+h_v}$), the corresponding optimization problems become more complex, since we must deal with the lengths of Jordan arcs on the surface $G_g$. Another variation of the foregoing problems is to require that the end points $x_o$ and $x_f$ belong to given disjoint subsets (e.g. nonintersecting curves) of $\Omega$. It will become evident later that some of the proposed numerical algorithms are also applicable to the modified problems with straightforward modifications.

## 4.1.1 Existence of Solutions

First, we consider Problem 4.1. The following result ensures that the set of all admissible paths $\Gamma$ satisfying condition (1) is nonempty.

**Proposition 4.1** *Under the conditions of Theorem 3.3, there exists an admissible path $\Gamma \in \Omega$ that satisfies condition (4.2).*

*Proof 4.1* From Theorem 3.3, there exists a finite point set $P^{(N)} = \{x^{(k)}, k = 1, \ldots, N\} \subset \Omega$ that satisfies the total visibility constraint (1). If the specified end points $x_o$ and $x_f \in P^{(N)}$, then under the assumption that $\Omega$ is simply connected, it is always possible to construct a path $\Gamma$ in $\Omega$ corresponding to a Jordan arc passing through all the points in $P^{(N)}$. In particular, if the line segment joining any pair of points in $P^{(N)}$ lies in $\Omega$, then a Jordan arc composed of straightline segments joining the successive points in $P^{(N)}$ can always be constructed. A trivial case is where $\Omega$ is a compact convex subset of $\mathbb{R}^2$. If $x_o$ and/or $x_f \notin P^{(N)}$, we augment $P^{(N)}$ by these points, and proceed with the construction of a Jordan arc. Finally, from the constructed $\Gamma$, the corresponding path $\Gamma_{\mathcal{P}}$ in $\mathcal{P}$ can be determined from $\{(x, g(x)) : x \in \Gamma\}$.                                                              $\square$

*Remark 4.3* Proposition 4.1 implies the existence of a Jordan arc passing through a finite set of observation points $(x^{(k)}, f_{h_v}(x^{(k)})) \in G_{f_{h_v}}$ such that $G_f$ is totally visible. In general, the set of observation points for total visibility is not unique. Moreover, the cardinality of this set depends on $f$.

*Remark 4.4* Suppose that the line segment $L$ joining the points $x_o$ and $x_f$ lies in $\Omega$, and $\bigcup_{x \in L} \mathcal{V}((x, f_{h_v}(x))) = G_f$, then $L$ is an optimal path. If $\bigcup_{x \in L} \mathcal{V}((x, f_{h_v}(x))) \subset G_f$, then for a certain class of $G_f$, the optimal path $\Gamma^*$ is close to $L$ in the sense of arc length. Minimal excursions from $L$ can be introduced so that the invisible part $G_f - \bigcup_{x \in L} \mathcal{V}((x, f_{h_v}(x)))$ becomes visible.

For Problem 4.1", we first construct the set $\tilde{\Omega} \stackrel{\text{def}}{=} \{x \in \Omega : \|\nabla f(x)\| \leq f'_{\max}\}$, and then find the shortest admissible path $\Gamma^* \subset \tilde{\Omega}$ such that $\bigcup_{x \in \Gamma^*} \mathcal{V}((x, f_{h_v}(x))) = G_f$. In general, it is possible that $\tilde{\Omega}$ consists of disjoint subsets of $\Omega$, and there may not exist admissible paths in $\tilde{\Omega}$ (e.g. $x_o$ and $x_f$ lie in two disjoint subsets of $\Omega$ separated by a strip on which $\|\nabla f(x)\| > f'_{\max}$ for all points $x$ on this strip). Consequently, Problem 4.1" has no solution. Note also that the line segment $L$ joining $x_o$ and $x_f$ may not lie in $\tilde{\Omega}$.

Since $f$ is a $C_1$-function, the set $\tilde{\Omega}$ is compact. Assuming the existence of an admissible path in $\tilde{\Omega}$, the observations given in Remarks 4.3 and 4.4 are also applicable to this case.

For Problem 4.2, we observe that in general, the set-valued mapping $x \rightarrow \Pi_\Omega \mathcal{V}((x, f_{h_v}(x)))$ on $\Omega$ into $2^\Omega$ may be discontinuous with respect to metrics $\rho_E$ and $\rho_H$. Consequently, the real-valued function $x \rightarrow \mu_2\{\Pi_\Omega \mathcal{V}((x, f_{h_v}(x)))\}$ on $\Omega$ into $\mathbb{R}$ may be discontinuous also. The following example illustrates this situation.

*Example 4.1* Let $\Omega$ be the unit disk $\{(r, \theta) : 0 \leq r \leq 1, 0 \leq \theta \leq 2\pi\}$ in $\mathbb{R}^2$. Consider the $C_1$-function $f$ given by

$$
f(r, \theta) = \begin{cases} 1 - \cos(2\pi r), & \text{if } 0 \leq r \leq 1/2; \\ 2, & \text{if } 1/2 \leq r \leq 1 \end{cases} \tag{4.7}
$$

defined on $\Omega$. The visible set corresponding to a point $(r, \theta, f_{h_v}(r, \theta)) \in G_{f_{h_v}}$, for $0 \leq h_v < 2, 0 \leq r \leq \bar{r}, 0 \leq \theta \leq 2\pi$, is given by

$$
\mathcal{V}((r, \theta, f_{h_v}(r, \theta))) = \begin{cases} \{(r', \theta', f(r', \theta')) : 0 \leq r' \leq \hat{r}_{\min}, 0 \leq \theta' \leq 2\pi\}, & \text{if } 0 \leq r < \bar{r}; \\ G_f, & \text{if } r = \bar{r}, \end{cases} \tag{4.8}
$$

where $\bar{r} = (\cos^{-1}(h_v - 1))/2\pi$, and $\hat{r}_{\min}$ is the smallest positive root of the following equation for $\hat{r}$:

$$
\cos(2\pi r) - \cos(2\pi \hat{r}) - h_v = 2\pi(\hat{r} - r)\sin(2\pi \hat{r}). \tag{4.9}
$$

The mapping $(r, \theta) \rightarrow \mathcal{V}((r, \theta, f_{h_v}(r, \theta)))$ is discontinuous at $(r, \theta) = (\bar{r}, \theta)$, $0 \leq \theta \leq 2\pi$. The discontinuity is induced by the flat portion of the surface $G_f$.

Now, if we assume that the real-valued function $x \rightarrow \mu_2\{\Pi_\Omega \mathcal{V}((x, f_{h_v}(x)))\}$ on $\Omega \rightarrow \mathbb{R}$ is $C_1$, then $G_{\mu_2}$ (graph of the mapping $\mu_2$) is a two-dimensional $C_1$-surface in $\mathbb{R}^3$. Consider the set $\tilde{A}_\Omega$ consisting of all admissible directed $C_1$-Jordan arcs $\tilde{\Gamma}$ connecting the end points $x_o$ and $x_f$ that satisfy the arc length constraint (4.4). Clearly, the function $x \rightarrow \mu_2\{\Pi_\Omega \mathcal{V}((x, f_{h_v}(x)))\}$ restricted to such an admissible arc $\tilde{\Gamma}$ is $C_1$ with respect to the arc length. Since $\mu_2\{\Pi_\Omega \mathcal{V}((x, f_{h_v}(x)))\} > 0$ for any $x \in \Omega$, it is evident that along any admissible path $\tilde{\Gamma} \in \tilde{A}_\Omega$, the line integral along $\tilde{\Gamma}$ given by

$$
\int_0^l \mu_2\{\Pi_\Omega \mathcal{V}((x, f_{h_v}(x))|_{x \in \tilde{\Gamma}})\} dl \tag{4.10}
$$

increases monotonically with the traversed arc length $l$ as the arc's end-point moves towards $x_f$. But in general, the length of the optimum path $\tilde{\Gamma}^*$ may not equal to $l_{\max}$. In fact, the maximum value of $J_1$ may correspond to directed admissible paths with the *shortest* length. The following example illustrates this situation.

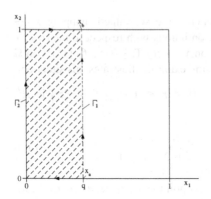 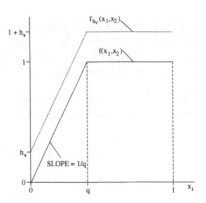

**Fig. 4.2**  Paths $\Gamma_1$ and $\Gamma_2$ for Example 4.2

*Example 4.2* Let $\Omega = \{(x_1, x_2) \in \mathbb{R}^2 : 0 \leq x_1 \leq 1, 0 \leq x_2 \leq 1\}$, and

$$f(x_1, x_2) = \begin{cases} x_1/q, & \text{if } 0 \leq x_1 \leq q; \\ 1, & \text{if } q \leq x_1 \leq 1, \end{cases} \tag{4.11}$$

where $0 < q < 1$. Let the end points be specified by $x_a = (q, 0)$ and $x_b = (q, 1)$. Consider two distinct admissible paths $\Gamma_1$ and $\Gamma_2$ connecting $x_a$ and $x_b$ as shown in Fig. 4.2. Let $h_v$ be a positive number $<1$. Then, along $\Gamma_1$,

$$\Pi_\Omega \mathcal{V}((q, x_2, 1 + h_v)) = \Omega \text{ for every } x_2 \in [0, 1].$$

Thus, $J_1(\Gamma_1) = \mu_2\{\Omega\} = 1$. Now, along $\Gamma_2$,

$$\Pi_\Omega \mathcal{V}((x_1, x_2, h_v)) = \begin{cases} \Omega_q \stackrel{\text{def}}{=} \{(x_1, x_2) \in \Omega : 0 \leq x_1 \leq q\} & \text{for } 0 \leq x_1 \leq (1 - h_v)q; \\ \Omega & \text{for } q(1 - h_v) \leq x_1 \leq q. \end{cases} \tag{4.12}$$

Evidently, $\mu_2\{\Omega_q\} = q$. Evaluating the integral in $J_1$ along $\Gamma_2$ gives

$$J_1(\Gamma_2) = 2q^2(1 - h_v) + q(1 + 2h_v). \tag{4.13}$$

It can be verified that there exists an $q \in ]0, 1[$ such that $J_1(\Gamma_1) > J_1(\Gamma_2)$ (e.g. for $h_v = 1/2, q = \sqrt{2} - 1$). But $\Lambda(\Gamma_2) > \Lambda(\Gamma_1) = 1$, where $\Gamma_1$ is the *shortest* admissible path connecting $x_a$ and $x_b$. Note that

$$0 < J_1(\Gamma) \leq l_{\max}\mu_2\{\Omega\} \tag{4.14}$$

for any admissible path $\Gamma \in \tilde{\mathcal{A}}_\Omega$. But the upper bound can be attained if and only if $G_f$ is totally visible from $(x, f_{h_v}(x))$ for every $x \in \Gamma$ such that $\Lambda(\Gamma) = l_{\max}$.

Although here $f$ is only piecewise $C_1$, we can approximate $f$ as close as we wish by a $C_1$-function, and obtain a similar result.

For Problem 4.2', we have $J_2(\Gamma) \le \mu_2(\Omega)$, and equality holds if there exists a finite point set $P^{(N)} = \{x^{(1)}, \ldots, x^{(N)}\} \subset \Omega$ with $x^{(1)} = x_a$, $x^{(N)} = x_b$ such that

$$\bigcup_{k=1}^{N} \Pi_\Omega \mathcal{V}((x^{(k)}, f_{h_v}(x^{(k)}))) = \Omega, \tag{4.15}$$

and a directed admissible Jordan curve connecting $x_a$ and $x_b$ that passes through all the remaining points of $P^{(N)}$, and satisfies the arc length constraint (4.4).

## 4.1.2 Optimality Conditions

Here, we develop optimality conditions for Problem 4.1 under the assumption that a solution exists. From Theorem 3.3, there exists a finite point set $P^{(N)} = \{x^{(k)}, k = 1, \ldots, N\} \subset \Omega$ with $x^{(1)} = x_o$ and $x^{(N)} = x_f$ such that

$$\bigcup_{k=1}^{N} \mathcal{V}((x^{(k)}, f_{h_v}(x^{(k)}))) = G_f \tag{4.16}$$

is satisfied. Let $P$ denote the set of all such $P^{(N)}$'s with $2 \le N < \infty$. The cardinality of $P$ is generally infinite. Now, for a given $P^{(N)} \subset P$, let $\tilde{\mathcal{A}}_{P^{(N)}}$ denote the set of all admissible paths $\Gamma$ formed by line segments joining distinct pairs of points in $P^{(N)}$. The cardinality of $\tilde{\mathcal{A}}_{P^{(N)}}$ is $\le (N-2)!$. Thus, Problem 4.1 reduces to finding a finite point set $P^{(N)} \subset P$ and an admissible path $\Gamma \in \tilde{\mathcal{A}}_{P^{(N)}}$ such that the arc length

$$\Lambda(\Gamma) = \sum_{k=1}^{N-1} \|x^{(k+1)} - x^{(k)}\|$$

is minimized. For convenience, the points $x^{(k)}, k = 2, \ldots, N-1$ are indexed consecutively along the path $\Gamma$ initiating from $x_o$ and moving towards the end point $x_f$.

The following simple necessary condition for optimality can be deduced readily from the definition of arc length:

**Proposition 4.2** *Let $\Gamma^*$ be an optimal admissible path for Problem 4.1, and $P^{(N^*)} = \{x_*^{(k)} \in \Gamma^*, k = 1, \ldots, N^*\}$ be a finite point set satisfying (4.16). Then, for any perturbed admissible path $\Gamma \in \tilde{\mathcal{A}}_{\tilde{P}^{(N^*)}}$ with finite point set $\{x_*^{(1)}, \ldots, x_*^{(k-1)}, x_*^{(k)} + \delta x, x_*^{(k+1)}, \ldots, x_*^{(N^*)}\}$ in which the point perturbation $\delta x$ about $x_*^{(k)}$ satisfies $x_*^{(k)} + \delta x \in \Omega$, and condition (4.16) holds, the following inequality:*

$$\|x_*^{(k)} + \delta x - x_*^{(k-1)}\| + \|x_*^{(k+1)} - x_*^{(k)} - \delta x\| \geq \|x_*^{(k+1)} - x_*^{(k)}\|$$
$$+ \|x_*^{(k)} - x_*^{(k-1)}\|, \quad k = 2, \ldots, N^* - 1, \tag{4.17}$$

*must be satisfied.*

For Problem 4.2, a necessary condition for optimality can be derived by considering local path perturbations. Let $I = [0, 1]$. First, we parameterize an admissible path $\Gamma$ by the scalar parameter $\lambda \in I$, i.e.

$$\Gamma = \{(x_1, x_2) \in \Omega : x_1 = q_1(\lambda), x_2 = q_2(\lambda), \lambda \in I\},$$

where $q_1$ and $q_2$ are real-valued $C_1$-functions on $I$ satisfying

$$(q_1(0), q_2(0)) = (x_{o1}, x_{o2}) \text{ and } (q_1(1), q_2(1)) = (x_{f1}, x_{f2}).$$

Let $\Gamma^*$ and $\Gamma$ denote an optimal path and an admissible perturbed path specified by

$$q^*(\lambda) = (q_1^*(\lambda), q_2^*(\lambda)) \text{ and } q^*(\lambda) + \eta(\lambda) = (q_1^*(\lambda) + \eta_1(\lambda), q_2^*(\lambda) + \eta_2(\lambda)), \quad \lambda \in I, \tag{4.18}$$

respectively, where $\eta = (\eta_1, \eta_2) \in \Sigma_{\Gamma^*}$, the set of all admissible path perturbations about $\Gamma^*$ defined by $\Sigma_{\Gamma^*} = \{\eta \in C_1(I; \mathbb{R}^2) : \eta(0) = (0, 0), \eta(1) = (0, 0); q^*(\lambda) + \eta(\lambda) \in \Omega \text{ for all } \lambda \in I\}$, with $C_1(I; \mathbb{R}^2)$ being the normed linear space of all continuous functions defined on $I$ with their values in $\mathbb{R}^2$ and having continuous first derivatives on $I$, and normed by: $\|\eta\|_m = \sum_{i=1,2}(\max_{\lambda \in I} |\eta_i(\lambda)| + \max_{\lambda \in I} |\eta_i'(\lambda)|)$, where $(\cdot)'$ denotes differentiation with respect to $\lambda$.

Consider the arc length corresponding to an admissible perturbed path $\Gamma$ given by

$$\Lambda(\Gamma) = \int_0^1 \sqrt{(q_1'^*(\lambda) + \eta_1'(\lambda))^2 + (q_2'^*(\lambda) + \eta_2'(\lambda))^2} d\lambda. \tag{4.19}$$

The arc length constraint (4.4) implies that $l_{\min} \leq \Lambda(\Gamma) \leq l_{\max}$.

Consider the increment in $J_1$ defined by

$$\Delta J_1 \stackrel{\text{def}}{=} J_1(\Gamma^*) - J_1(\Gamma) = \int_0^1 (\mu_n\{S(q^*(\lambda))\} - \mu_n\{S(q^*(\lambda)) + \eta(\lambda)\}) d\lambda, \tag{4.20}$$

where $S(q^*(\lambda)) = \Pi_\Omega V((q^*(\lambda), f_{h_v}(q^*(\lambda))))$. To proceed further, we make use of the following elementary identity:

**Fig. 4.3** Illustration of set identities in (4.22)

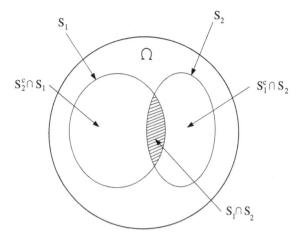

**Lemma 4.1** *Let $S_1$ and $S_2$ be connected measurable subsets of $\Omega \subset \mathbb{R}^n$. Then,*

$$\mu_n\{S_1\} - \mu_n\{S_2\} = \mu_n\{S_1 \cap S_2^c\} - \mu_n\{S_1^c \cap S_2\}, \qquad (4.21)$$

*where $S_i^c$ denotes the complement of $S_i$ relative to $\Omega$.*

*Proof 4.2* From the following set identities (see Fig. 4.3 for a sketch):

$$S_1 = \{S_1 \cap S_2^c\} \cup \{S_1 \cap S_2\}, \ S_2 = \{S_1^c \cap S_2\} \cup \{S_1 \cap S_2\}, \qquad (4.22)$$

we have

$$\mu_n\{S_1\} = \mu_n\{S_1 \cap S_2^c\} + \mu_n\{S_1 \cap S_2\}, \quad \mu_n\{S_2\} = \mu_n\{S_1^c \cap S_2\} + \mu_n\{S_1 \cap S_2\}, \qquad (4.23)$$

and identity (4.21) follows. $\qquad \square$

Setting $S_1 = S(q^*(\lambda))$ and $S_2 = (q^*(\lambda) + \eta(\lambda))$ in Lemma 4.1, the integrand of the integral in (4.20) can be rewritten as:

$$\mu_n\{S(q^*(\lambda))\} - \mu_n\{S(q^*(\lambda) + \eta(\lambda))\} = \mu_n\{S(q^*(\lambda)) \cap S(q^*(\lambda) + \eta(\lambda))^c\}$$
$$- \mu_n\{S(q^*(\lambda))^c \cap S(q^*(\lambda) + \eta(\lambda))\}, \qquad (4.24)$$

where $S^c$ denotes the complement of $S$ relative to $\Omega$. Evidently, a sufficient but not necessary condition for $\tilde{\Gamma}^*$ satisfying the arc length constraint (4.4) to be optimal is that

$$\mu_n\{S(q^*(\lambda)) \cap S(q^*(\lambda) + \eta(\lambda))^c\} \geq \mu_n\{S(q^*(\lambda))^c \cap S(q^*(\lambda) + \eta(\lambda))\} \quad (4.25)$$

for all $0 < \lambda < 1$, and all $\eta \in \Sigma_{\Gamma^*}$ such that the perturbed path $\Gamma$ corresponding to $q^* + \eta$ satisfies the arc length constraint (4.4).

Now, since $\tilde{\Gamma}^*$ is optimal, $\Delta J_1 \geq 0$ for all admissible $\tilde{\Gamma}$ satisfying (4.4). Thus we have the following necessary condition for optimality:

**Theorem 4.1** *Suppose that the function* $x \to \mu_n\{\Pi_\Omega V((x, f_{h_v}(x)))\}$ *restricted to the optimal path* $\Gamma^*$ *is* $C_1$. *Then the Gateaux differential of* $J_1$ *at* $\Gamma^*$ *with increment* $\eta$ *exists and satisfies:*

$$DJ_1(\Gamma^*; \eta) = \int\limits_0^1 \lim_{\alpha \to 0} \frac{1}{\alpha} (\mu_n\{S(q^*(\lambda)) \cap S(q^*(\lambda) + \alpha\eta(\lambda))^c\}$$

$$- \mu_n\{S(q^*(\lambda))^c \cap S(q^*(\lambda) + \alpha\eta(\lambda))\})d\lambda \geq 0 \qquad (4.26)$$

*for all* $\eta \in \Sigma_{\Gamma^*}$, *where* $S(q^*(\lambda)) = \Pi_\Omega V((q^*(\lambda), f_{h_v}(q^*(\lambda))))$.

Due to the complex relation between the observation point and its visible set, the Gateaux differential in (4.26) cannot be easily computed except for simple cases. In what follows, we shall consider approximations of Problems 4.1 and 4.2, and develop numerical algorithms for their solutions.

### 4.1.3 Numerical Algorithms

To facilitate the development of numerical algorithms for the optimal path planning problems, a mesh on $\Omega$ using standard methods such as Delaunay triangulation is established. Then $G_f$ is approximated by a polyhedral surface $\hat{G}_f$, in particular, a surface formed by triangular patches. In practical situations, the function $f = f(x)$ is usually given in the form of numerical data. An approximation of $G_f$ can be obtained by interpolation of the given numerical data. Here, algorithms are developed for the approximate Problems 4.1 and 4.2 that make use of the numerical data directly.

For the Shortest Path Problem 4.1, consider the simplest case where the line segment $L$ joining the end points $x_o$ and $x_f$ lies in $\Omega$. Remark 4.4 suggests that a possible approach to obtaining a solution to the approximate Problem 4.1 is to seek first a finite number of points along $L$ having maximal visibility, and then introduce additional points close to $L$ to achieve total visibility. Finally, a Jordan-arc with minimal length passing through all the observation points is constructed.

Suppose that on $\Omega$, a mesh consisting of points $x^{(k)}, k = 1, \ldots, M$ along with the approximate surface $\hat{G}_f$ formed by triangular patches have been established. For convenience, let $x^{(1)} = x_o$ and $x^{(M)} = x_f$. Let $\hat{\mathcal{G}}$ denote the set of all Jordan arcs connecting $x^{(1)}$ and $x^{(M)}$ formed by line segments corresponding to the edges of the triangles. The basic steps in our algorithm for determining an optimal path for the approximate Problem 4.1 without assuming that the line segment $L$ joining $x_o$ and $x_f$ lies in $\Omega$ are as follows:

**Step 1** Construct the set $\hat{\mathcal{G}}$.
**Step 2** Partially order $\hat{\mathcal{G}}$ according to the arc length.

**Step 3** Determine $\hat{\mathcal{G}}^* = \{\Gamma_j^*, j = 1, \ldots, K\}$, the set of all Jordan arcs in $\hat{\mathcal{G}}$ having the shortest path length.

**Step 4** Select a path in $\hat{\mathcal{G}}^*$, say $\Gamma_i^*$.

**Step 5** Compute the visible set $\mathcal{V}((x^{(k)}, f_{h_v}(x^{(k)})))$ corresponding to each point $x^{(k)}$ along the path $\Gamma_i^*$ and its projection $\Pi_\Omega \mathcal{V}((x^{(k)}, f_{h_v}(x^{(k)})))$.

**Step 6** Determine whether there exists a combination of points $x^{(k)}$ along $\Gamma_i^*$ such that the union of their visible sets $\mathcal{V}((x^{(k)}, f_{h_v}(x^{(k)})))$ is equal to $G_f$.

If YES, then $\Gamma_i^*$ is an optimal path for the approximate Problem 4.1, STOP;

if NO, select a neighboring path of $\Gamma_i^*$ in the sense of arc length, and GO TO Step 5.

*Remark 4.5* In Steps 2 and 3, instead of considering triangles in $\Omega$, we may consider triangular patches associated with the polyhedral approximation of the surface $G_f$. Efficient algorithms for computing the shortest arc length such as those due to Sharir and Shorr [3], O'Rourke et al. [4], and Lawler [5] may be used.

*Remark 4.6* Step 5 involves the computation of visible sets $\mathcal{V}((x^{(k)}, f_{h_v}(x^{(k)})))$ associated with points $x^{(k)}$ along the path $\Gamma_i^*$, an NP-hard problem in computational geometry. This task can be accomplished using a suitable algorithm developed recently by Balmes and Wang [1]. The complexity of that algorithm is $O(np^2)$, where $n$ and $p$ are the number of observation points and the number of triangular patches respectively. This algorithm can be easily modified to take into account the limited aperture of cameras or sensors.

The foregoing algorithm can be modified to solve Problem 4.1″. As before, we introduce a mesh on $\Omega$ and obtain an approximate surface $\hat{G}_f$. Next, we compute the approximate gradient of $f$ from the values of $f$ at the mesh points, and then delete all those mesh points at which the gradient constraint (4.3) is violated. Thus, we obtain a mesh for the spatial domain $\tilde{\Omega}$ as defined in Sect. 4.1. Assuming that $\tilde{\Omega}$ is a simply connected set in $\mathbb{R}^2$, we can construct the set $\hat{\mathcal{G}}$ in $\tilde{\Omega}$, and follow the steps of the algorithm as before.

Now, we consider the Maximum Visibility Problem 4.2. As in Problem 4.1, we first establish a mesh $x^{(k)}, k = 1, \ldots, M$ on $\Omega$, and approximate $G_f$ by a polyhedral surface $\hat{G}_f$ formed by triangular patches. A numerical algorithm for the approximate Problem 4.2 is given by the following basic steps:

**Step 1** Compute the measure $\mu$ of the projection of visible set on $\Omega$ for each point $(x^{(k)}, f_{h_v}(x^{(k)})) \in \hat{G}_{f_{h_v}}$ (using an algorithm such as that of Balmes and Wang [1]), and construct the corresponding polyhedral surface $G_\mu$.

**Step 2** Find a Jordan arc $\Gamma \subset \Omega$ with end points $x^{(1)} = x_o$ and $x^{(M)} = x_f$, and formed by line segments passing through the points $x^{(k)}, k \in \mathcal{I} \subseteq \{2, \ldots, M-1\}$, that satisfies the arc length constraint $l_{\min} \leq \Lambda(\Gamma) \leq l_{\max}$ such that

$$\hat{J}_1(\Gamma) \overset{\text{def}}{=} \sum_{k \in \mathcal{I} \cup \{1, M\}} \mu_n\{\Pi_\Omega \mathcal{V}((x^{(k)}, f_{h_v}(x^{(k)})))\} \tag{4.27}$$

takes on its maximum value.

*Remark 4.7* The combinatorial optimization problem in Step 2 is akin to the "shortest path problem" in networks [5]. Here, the admissible paths connecting points $x_o$ and $x_f$ are represented by a directed graph whose nodes correspond to the mesh points, and the path lengths satisfy the arc length constraint. Moreover, we only consider directed paths from $x_o$ to $x_f$ without cycles or loops. The "weight" associated with the arc connecting a pair of adjacent nodes or mesh points is specified by the measure of the visibility set of the arc end-point. With the foregoing modification, either Dijkstra [2] or dynamic programming method may be used to obtain numerical solutions to the problem.

The foregoing algorithm for Problem 4.2 can be modified for solving Problem 4.2'. In Step 1, we only need to compute the visible set for each point in $G_{f_{h_v}}$. In Step 2, we replace $\hat{J}_1(\Gamma)$ by

$$\hat{J}_2(\Gamma) \overset{\text{def}}{=} \mu_n \Big\{ \bigcup_{k \in \mathcal{I} \cup \{1, M\}} \Pi_\Omega \mathcal{V}((x^{(k)}, f_{h_v}(x^{(k)}))) \Big\}. \tag{4.28}$$

### 4.1.4 Numerical Examples

First, consider a simple example for which the visible sets at any point in $G_{f_{h_v}}$ can be computed analytically.

*Example 4.3* Let the spatial domain be the unit disk in $\mathbb{R}^2$ given by $\Omega = \{(x_1, x_2) : x_1^2 + x_2^2 \leq 1\}$. Let the surface under observation be specified by the graph of $f$ given by

$$f(x_1, x_2) = 1 - x_1^2 - x_2^2, \tag{4.29}$$

defined on $\Omega$. Elementary computations show that for any given $h_v > 0$, the projection of the visible set from a point $(x_1, x_2, f_{h_v}(x_1, x_2)) \in G_{f_{h_v}}$ on $\Omega$ is simply the intersection of the unit disk with the disk centered at $(x_1, x_2)$ with radius $\sqrt{h_v}$, i.e.

$$\Pi_\Omega \mathcal{V}((x_1, x_2, f_{h_v}(x_1, x_2))) = \Omega \cap \{(x_1', x_2') \in \mathbb{R}^2 : (x_1' - x_1)^2 + (x_2' - x_2)^2 \leq h_v\}. \tag{4.30}$$

It can be verified that for $0 < h_v \leq 1$, the measure of $\Pi_\Omega \mathcal{V}((x_1, x_2, f_{h_v}(x_1, x_2)))$ is given by

$$\mu_2\{\Pi_\Omega \mathcal{V}((x_1, x_2, f_{h_v}(x_1, x_2)))\} = \begin{cases} h_v \pi, & \text{if } 0 \leq r \leq 1 - \sqrt{h_v}; \\ \kappa(r, \tilde{r}), & \text{if } 1 - \sqrt{h_v} < r \leq 1; \end{cases} \tag{4.31}$$

and for $h_v \geq 1$,

$$\mu_2\{\Pi_\Omega \mathcal{V}((x_1, x_2, f_{h_v}(x_1, x_2)))\} = \begin{cases} \pi, & \text{if } 0 \leq r \leq \sqrt{h_v} - 1; \\ \kappa(r, \tilde{r}), & \text{if } \sqrt{h_v} - 1 < r \leq 1, \end{cases} \quad (4.32)$$

where $r = \sqrt{x_1^2 + x_2^2}$, $\tilde{r} = (1 - h_v + r^2)/2r$, and

$$\kappa(r, \tilde{r}) = \cos^{-1}(\tilde{r}) - \tilde{r}\sqrt{1 - \tilde{r}^2} + h_v \cos^{-1}\left(\frac{r - \tilde{r}}{\sqrt{h_v}}\right) - (r - \tilde{r})\sqrt{h_v - (r - \tilde{r})^2}.$$
$$(4.33)$$

Figure 4.4 shows the surface corresponding to (4.29). Figure 4.5a, b show the surfaces corresponding to (4.31) and (4.32) for $h_v = 1$ and $h_v = 2$ respectively. Note that for $h_v = 1$, the surface $G_{\mu_2}$ is not smooth at the origin. Moreover, $h_v = 1$ corresponds to the critical height $h_{vc}$ at the origin as defined in Proposition 2.1.

Now, let

$$x_o = (-1, 0) \text{ and } x_f = (1, 0).$$

Consider the Shortest Path Problem 4.1. Evidently, for $h_v \geq 1$, the optimal path $\Gamma^*$ is the line segment joining $x_o$ and $x_f$, since at the origin

$$x = (0, 0), \quad \Pi_\Omega \mathcal{V}((0, 0, f_{h_v}(0, 0))) = \Omega.$$

For $0 < h_v \leq 1$, the admissible paths $\Gamma$ that satisfy the total visibility condition (4.2) can be determined by using the procedure outlined in the proof of

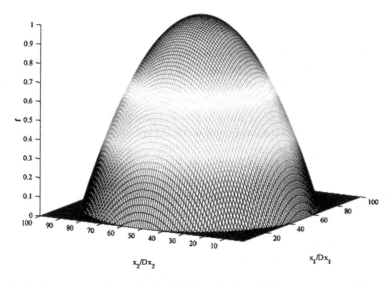

**Fig. 4.4** Surface corresponding to Example 4.3 described by (4.29); $\Delta x_1 = \Delta x_2 = 0.02$

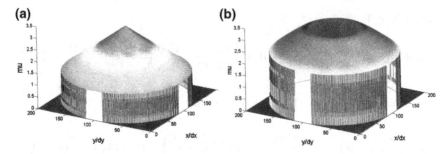

**Fig. 4.5** Measures of visible sets corresponding to Example 4.3 described by (4.31) with $h_v = 1$ (Fig. (a)), and (4.32) with $h_v = 2$ (Fig. (b)); $\Delta x_1 = \Delta x_2 = 0.02$

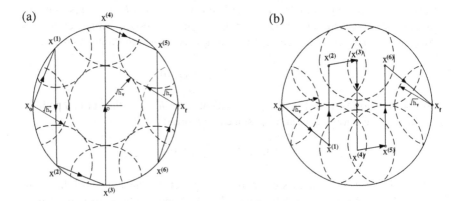

**Fig. 4.6** Construction of admissible paths $\Gamma$ in Problem 4.2, Example 4.3 for total visibility

Proposition 4.1. First, we determine $\Pi_\Omega \mathcal{V}((-1, 0, f_{h_v}(-1, 0)))$ and $\Pi_\Omega \mathcal{V}((0, 1, f_{h_v}(0, 1)))$ by constructing the circular disks with radius $\sqrt{h_v}$ centered at $x_o$ and $x_f$ respectively. Next, we cover the boundary of $\Omega$ by selecting a finite set of points $x^{(1)}, \ldots, x^{(6)}$ on the boundary as shown in Fig. 4.6a. If we construct the path $\Gamma$ from $x_o$ to $x_f$ by connecting $x_o, x^{(1)}, \ldots, x^{(6)}, x_f$ consecutively via line segments, then $G_f$ is totally visible from $G_{f_{h_v}}$ restricted to $\Gamma$. Clearly, the path $\Gamma$ connecting $x_o$ and $x_f$ that satisfies total visibility condition (4.16) is non-unique. Figure 4.6b shows another admissible path having the same visibility property. It can be verified that the solution to Problem 4.1 can be determined by first finding the minimum number of circular disks with radius $\sqrt{h_v}$ and centered at $x^{(k)}, k = 1, \ldots, N$, and at an equal distance from the origin such that the union of these disks covers the boundary. Then the optimal path is the one that connects the points $x_o, x^{(1)}, \ldots, x^{(N)}, x_f$ by line segments with minimum length. The path shown in Fig. 4.6b corresponds to an optimal path for $h_v = 1/2$.

Next, we consider the Maximum Visibility Problem 4.2 with $l_{\min} = 2$. For the trivial case where $l_{\max} = 2$ and $h_v \geq 1$, the optimal path $\Gamma^*$ is simply the line segment connecting $x_o = (-1, 0)$ and $x_f = (1, 0)$. For the case where $0 < h_v < 1$

**Fig. 4.7** Optimal path $\Gamma^*$
for Problem 4.2, Example 4.3
with $0 < h_v < 1$ and $l_{max} > 2$

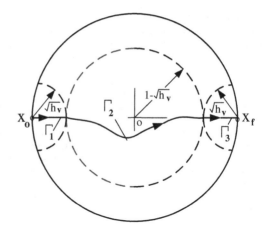

and $l_{max} > 2$, the solution to Problem 4.2 is evident from Fig. 4.7. At $x_o = (-1, 0)$ and $x_f = (1, 0)$, the corresponding projections of the visible sets on $\Omega$ are the intersections of the unit disk and the disks with radius $\sqrt{h_v}$ centered at $x_o$ and $x_f$ respectively. For any point $x$ in the disk $D_o$ centered at the origin with radius $1 - \sqrt{h_v}$, we have $\mu\{\Pi_\Omega \mathcal{V}((x, f_{h_v}(x)))\} = h_v \pi$. Clearly, the maximum value of $J$ is attained by a directed admissible path $\Gamma^* = \bigcup_{i=1}^{3} \Gamma_i$, where $\Gamma_1$ and $\Gamma_3$ are the line segments on the $x_1$-axis (shortest arcs) connecting $x_o$ and $x_f$ with disk $D_o$ respectively. $\Gamma_2$ is any directed Jordan arc in $\Omega$ with length $l_{max} - 2\sqrt{h_v}$ that initiates from $(-1 + \sqrt{h_v}, 0)$ and terminates at $(1 - \sqrt{h_v}, 0)$.

*Example 4.4* Consider the optimal path planning problem for the simulated Mars surface with a small crater inside a large one discussed in Chap. 3, Sect. 3.4, Example 3.7. The surface $\hat{G}_f$ and its level curves are shown in Fig. 4.8a, b respectively. Assuming that the cameras/sensors are attached to a platform with fixed vertical height $h_v$, the observation platform corresponds to the elevated surface $\hat{G}_{f_{h_v}}$.

**(a)**　　　　　　　　　　　**(b)**

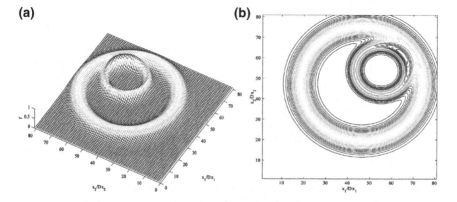

**Fig. 4.8** **a** Mars surface model. **b** Level contours of the Mars surface model; $\Delta x_1 = \Delta x_2 = 0.25$ km

**Fig. 4.9** Approximation of
the simulated Mars surface by
triangular patches

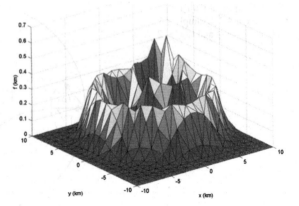

Here, the visible set of a point on the surface $\hat{G}_{f_{h_v}}$ is a subset of $\hat{G}_f$. Figure 4.9 shows the simulated Mars surface approximated by triangular patches. The visible and invisible sets of the initial observation point $z_o = (-10, -10, 0.002)$ km for the simulated Mars surface and their projections on $\Omega$ are shown in Fig. 4.10. Using the algorithm of Balmes and Wang [1] and a $51 \times 51$ mesh, $\mu\{\Pi_\Omega \mathcal{V}((x, f_{h_v}(x)))\}$, the measure of projections of visible sets on the spatial domain $\Omega$, is computed for each mesh point $x$ and $h_v = 0.002$ km. The results are shown in Fig. 4.11. Then a solution to the approximate Problem 4.1 with end points $x_o = (-10, -10)$ km and $x_f = (10, 10)$ km on a $21 \times 21$ mesh is determined numerically using the algorithm described in Sect. 4.1.3. The level curves corresponding to $\mu_2\{\Pi_\Omega \mathcal{V}(\cdot)\}$ and the computed optimal path are shown in Fig. 4.12. We observe that although the straightline $L$ connecting $x_o$ and $x_f$ lies in $\Omega$, the optimal path $\Gamma^*$ deviates significantly from $L$ due to the structure of $\hat{G}_f$. Moreover, $\Gamma^*$ traverses those mesh points with high visibility measures.

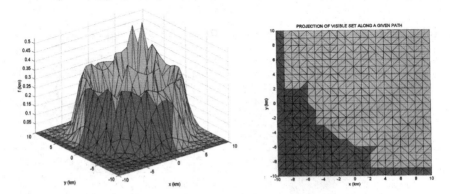

**Fig. 4.10** Visible set (*dark grey*) and invisible set (*light grey*) of the initial observation point $z_o = (-10, -10, 0.002)$ km for the simulated Mars surface and their projections on $\Omega$

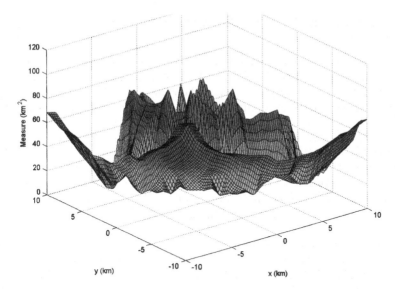

**Fig. 4.11** Measures of the projection of visible sets on $\Omega$ for the simulated Mars surface

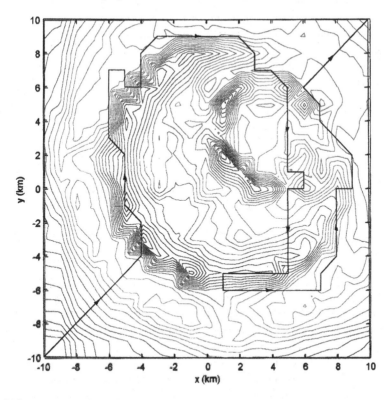

**Fig. 4.12** Level curves for $\mu_2\{\Pi_\Omega \mathcal{V}((\mathbf{x}, f_{h_v}(\mathbf{x})))\}$ for simulated Mars surface and an optimal path for approximate Problem 4.1

## 4.2 Observation Of Three-Dimensional Objects

Now, we turn our attention to optimal path planning problems for observing three-dimensional objects. As we mentioned in Chap. 2 (Proposition 2.2 and Remark 2.2) that at least two point-observers are required for total visibility of a solid body in $\mathbb{R}^3$. The following lemma gives a sufficient condition for total visibility of a solid body in $\mathbb{R}^3$.

**Lemma 4.2** *Suppose that the object under observation is a compact simply-connected solid body $\partial \mathcal{O} \subset \mathbb{R}^3$ with a smooth boundary $\partial \mathcal{O}$ and no outward-normal intersection points. Then, total visibility of $\mathcal{O}$ can be attained by a finite set of observation points in the observation platform $\mathcal{P}_h$ at a constant finite height $h$ along the outward normal above the body surface $\partial \mathcal{O}$.*

*Proof 4.3* Since $\mathcal{O}$ is a compact simply-connected solid body with a smooth boundary $\partial \mathcal{O}$ and no outward-normal intersection points, then $\mathcal{P}_h = \{z \in \mathbb{R}^3 : z = x + hn(x), x \in \partial \mathcal{O}\}$ is well-defined. For any observation point $z \in \mathcal{P}_h$, there exists a point $x = z - hn(x) \in \partial \mathcal{O}$ such that $x$ is visible from $z$. Moreover, $\mathcal{V}(z)$, the visible set of $z$, is a compact subset of $\partial \mathcal{O}$ containing $x$ with finite measure $\mu_2(\partial \mathcal{V}(z))$. Since $\mu_2(\mathcal{O})$ is finite, there exists a finite number $N$ of observation points $z^{(i)} \in \mathcal{P}_h, i = 1, \ldots, N$ such that $\partial \mathcal{O} \subseteq \bigcup_{i=1}^{N} \mathcal{V}(z^{(i)})$ as guaranteed by the Heine-Borel Covering Theorem.                                                                          □

First, we extend Problem 4.1, the shortest path total visibility problem, to the observation of a 3-D object $\mathcal{O}$ by a single mobile-observer.

**Problem 4.1A  Single Mobile-Observer Shortest Path Problem.** Given an observation platform $\mathcal{P}$ enclosing the observed 3-D object $\mathcal{O}$, and two distinct points $z_o, z_f \in \mathcal{P}$, find the shortest admissible path $\Gamma_{\mathcal{P}}$ starting at $z_o$ and ending at $z_f$ such that $\bigcup_{z \in \Gamma_{\mathcal{P}}} \mathcal{V}(z) = \partial \mathcal{O}$.

We observe that for $\mathcal{P} = \mathcal{P}_h$, once a solution to Problem 3.5 is obtained, any admissible path starting at $z_o$, passing through all the points in the observation-point set $\mathcal{P}^{(N)}$, and ending at $z_f$, is a candidate to the solution of Problem 4.1A.

To illustrate the nature of the solution to Problem 4.1A, we consider a specific example.

*Example 4.5* Here, the object under observation is a closed spherical ball $\bar{B}(0; r_o)$. The observation platform $\mathcal{P}_h$ is a sphere $\mathcal{S}(0; r_o + h)$ with finite observation height $h > 0$. From the Principle of Optimality in Dynamic Programming, an optimal path is a concatenation of geodesics starting and ending at distinct specified points $z_o, z_f \in \mathcal{P}_h$. Figure 4.13 shows an optimal path corresponding to $z_o = (0, -(r_o + h), 0)$ and $z_f = (0, r_o + h, 0)$, with observation height $h$ satisfying $0 < \alpha \leq \pi/4$, where $\alpha = \sin^{-1}(r_o/(r_o + h))$ is the half-angle of the observation cone. We observe that the optimal path has length $l_{min} = 3\pi(r_o + h)$, and has a great-circle loop along the path. For $\alpha > \pi/4$, there are more than one great-circle loops along the path.

**Fig. 4.13** An optimal path for Example 4.5

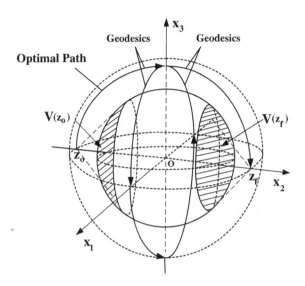

Moreover, the optimal path is non-unique. In fact, there are uncountably infinite number of optimal paths with length $l_{min}$.

Next, we generalize the Maximum Visibility Problem 4.2 to the observation of a 3-D object by a single mobile-observer.

**Problem 4.2A  Single Mobile-Observer Maximum Visibility Problem.** Given an observation platform $\mathcal{P}$ enclosing the observed 3-D object $\partial\mathcal{O}$, and two distinct points $z_o, z_f \in \mathcal{P}$, let $\tilde{\mathcal{A}}_\mathcal{P}$ denote the set of all admissible paths $\Gamma_\mathcal{P}$ from $z_o$ to $z_f$ satisfying the arc length constraint: $0 < l_{\min} \leq \mathcal{L}(\Gamma_\mathcal{P}) \leq l_{\max}$, where $\mathcal{L}(\Gamma_p)$ denotes the length of path $\Gamma_\mathcal{P}$. Find an $\Gamma_\mathcal{P}^* \in \tilde{\mathcal{A}}_\mathcal{P}$ such that the path visibility functional defined by the path integral:

$$J_1(\Gamma) = \int_{\Gamma_\mathcal{P}} \mu_2(\mathcal{V}(z))dl \qquad (4.34)$$

satisfies $J_1(\Gamma_\mathcal{P}^*) \geq J_1(\Gamma_\mathcal{P})$ for all $\Gamma_\mathcal{P} \in \tilde{\mathcal{A}}$, where $dl$ is the elemental arc length, and $\mu(\mathcal{V}(z))$ denotes the surface area of $\mathcal{V}(z)$.

It is convenient to parameterize an admissible path $\Gamma_\mathcal{P}$ by a scalar parameter $\lambda \in I = [0, 1]$, i.e. $\Gamma_\mathcal{P} = \{z \in \tilde{\mathcal{A}}_\mathcal{P} : z = q(\lambda), \lambda \in I\}$, where $q = (q_1(\lambda), q_2(\lambda), q_3(\lambda))$ with $q_i$ being real-valued smooth functions on $I$ satisfying $q(0) = z_o$ and $q(1) = z_f$.

**Theorem 4.2** *Suppose that an optimal path $\Gamma_\mathcal{P}^*$ exists, and the function $z \to \mu_2(\mathcal{V}(z))$ restricted to $\Gamma_\mathcal{P}^*$ is $C_1$. Then the Gateaux-differential of $J_1$ at $\Gamma_\mathcal{P}^*$ with increment $\eta$ exists and satisfies*

$$DJ_1(\Gamma_{\mathcal{P}}^*; \eta) = \int\limits_0^1 \lim_{\alpha \to 0} (1/\alpha)[\mu_2\{\mathcal{V}(q^*(\lambda) + \alpha\eta(\lambda))^c \cap \mathcal{V}(q^*(\lambda))\}$$

$$- \mu_2\{\mathcal{V}(q^*(\lambda))^c \cap \mathcal{V}(q^*(\lambda) + \alpha\eta(\lambda))\}]d\lambda \ge 0 \qquad (4.35)$$

*for all admissible* $\eta \in \Sigma_{\mathcal{A}^*} \overset{\text{def}}{=} \{\eta \in C_1(I; \mathbb{R}^3) : \eta(0) = 0, \eta(1) = 0; q^*(\lambda) + \alpha\eta(\lambda) \in \mathcal{P}$ *for all* $\lambda \in I\}, 0 \le \alpha < 1$, *where* $\mathcal{V}^c$ *denotes the complement of* $\mathcal{V}$ *relative to* $\partial\mathcal{O}$.

## Exercises

**Ex. 4.1** The object under observation consists of two opaque circular disks $D_1$ and $D_2$ in $\mathbb{R}^2$ centered at $(x_1, x_2) = (-3/2, 0)$ and $(3/2, 0)$ with radii $R_1$ and $R_2$ respectively as shown in Fig. 4.14. The observations are made by a point observer located on the line segment $\Gamma = [(-2, 0), (2, 0)]$.

(a) Assume $R_1 = R_2 = 1$. Determine the visible sets and their measures $\mu_1\{\mathcal{V}((x_1, 0))\}$ for all $(x_1, 0) \in \Gamma$. Is $\mu_1$ a smooth function of $x_1$?

(b) Repeat part (a) for $R_1 = 1$ and $R_2 = 1/2$.

**Ex. 4.2** Let $x = (x_1, x_2)$, $\Omega_1 = \{x \in \mathbb{R}^2 : (x_1 + 1)^2 + x_2^2 \le 1\}$, $\Omega_2 = \{x \in \mathbb{R}^2 : (x_1 - 1)^2 + x_2^2 \le 1\}$, and $\Omega = \Omega_1 \cup \Omega_2$. Let $f_1(x) = 1 - (x_1 + 1)^2 - x_2^2$, $f_2(x) = 1 - (x_1 - 1)^2 - x_2^2$, and

$$f(x) = \begin{cases} f_1(x) & \text{if } x \in \Omega_1; \\ f_2(x) & \text{if } x \in \Omega_2. \end{cases} \qquad (4.36)$$

The object under observation corresponds to the surface in $\mathbb{R}^3$ described by $G_f$, the graph of $f$. The observation is made from the constant vertical-height observation platform $\mathcal{P}_{h_v} = G_{f+h_v}$, where $h_v > 0$ is the vertical height.

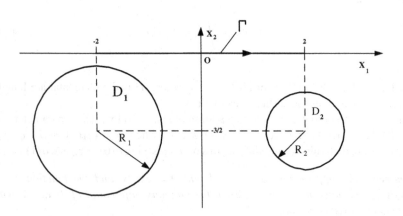

**Fig. 4.14** Object under observation in Problem 4.1A

(a) Examine the smoothness of $\mathcal{P}_{h_v}$ as the vertical height $h_v$ is varied.

(b) Find an observer point $(x, f(x) + h_v) \in \mathcal{P}_{h_v}$ for some $h_v > 0$ such that $G_f$ is totally visible from $(x, f(x) + h_v)$.

(c) Determine the shortest path $\Gamma$ connecting the initial point $z_o = (-2, 0, h_v)$ and final point $z_f = (2, 0, h_v)$ such that its corresponding visible set $\mathcal{V}(\Gamma) = G_f$. Is the shortest path for total visibility unique?

(Note $G_{f_1 + h_v} \cap G_{f_2 + h_v} = \{(0, 0, h_v)\}$. The observation path $\Gamma$ must lie in $\Omega$ or passes through the point $(0, 0, h_v)$).

# References

1. C.S. Balmes, P.K.C. Wang, Numerical algorithms for optimal visibility problems, UCLA Engineering Report ENG 00–214, June 2000
2. E.W. Dijkstra, A note on two problems in connection with graphs. Numerische Mathematik **1**, 269–271 (1959)
3. M. Sharir, A. Shorr, On shortest paths in polyhedral spaces. SIAM J. Comput. **15**, 193–215 (1986)
4. J. O'Rourke, S. Suri, H. Boot, Shortest paths on polyhedral surfaces, in *Proceedings of Symposium on Theoretical Aspects of Computer Science*, (1985) pp. 243–254
5. E.L. Lawler, *Combinatorial Optimization: Networks and Matroids* (Holt, Rinehart and Winston, New York, 1976)

# Chapter 5
# Visibility-based Optimal Motion Planning

So far, we have considered various visibility-based optimization problems in which either the point-observers are stationary or free to move along certain admissible paths. Their movements do not involve time and dynamical effects. The visibility-based optimal path planning problems under consideration in this chapter involve the dynamics of the observers. Therefore we refer to these problems as "visibility-based optimal *motion* planning problems". In physical situations, these observers may correspond to mobile robots, spacecraft or surveillance vehicles equipped with on-board cameras and/or sensors. To introduce the basic ideas and clarify problem formulations, we shall begin with a detailed discussion of a particular case in which the object under observation is a two-dimensional surface in $\mathbb{R}^3$, and there is only a single mobile point-observer whose motion (governed by Newton's Law) is confined to the observed surface (see Fig. 5.1). The observations are made from a constant vertical-height platform above the observed surface. Various optimal motion-planning problems for determining the mobile observer's motion or space-time trajectory whose projection onto the observation platform gives maximum visual coverage in some sense are posed as optimal control problems involving set measures. The conditions for the existence of solutions and their optimality for these problems in the form of variational inequalities and maximum principles will be developed along with algorithms for their numerical solutions. Their application will be illustrated by a simple example involving the observation of a simple surface in $\mathbb{R}^3$. Next, generalizations of the abovementioned visibility-based optimal motion planning-problems to optimal control problems involving set measures for a certain class of nonlinear dynamical systems will be considered. The results for the generalized problem is applied to the asteroid observation problem discussed in Chap. 1. We shall conclude this chapter with a discussion of real-time observer motion-planning problem with available surface data limited to those acquired from past and present observations.

© Springer International Publishing Switzerland 2015
P.K.-C. Wang, *Visibility-based Optimal Path and Motion Planning*,
Studies in Computational Intelligence 568, DOI 10.1007/978-3-319-09779-4_5

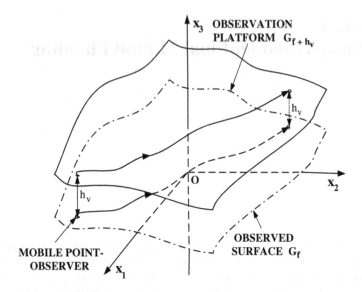

**Fig. 5.1** Observation of a surface by a mobile point-observer moving on a constant vertical-height platform

## 5.1 Particular Case

Let $B = \{e_1, \ldots, e_n\}$ be a specified orthonormal bases for the $n$-dimensional real Euclidean space $\mathbb{R}^n$, where $e_i$ corresponds to the $i$th unit basis vector. The representation of a point $x \in \mathbb{R}^n$ with respect to $B$ is specified by the column vector $[x_1, \ldots, x_n]^T$. The usual Euclidean norm of $x$ is denoted by $\|x\|$. Let $f = f(x)$ be a specified real-valued $C_2$-function defined on $\Omega$, a specified simply connected, compact subset of $\mathbb{R}^2$ with a smooth boundary $\partial\Omega$. As in Chap. 2, the graph and epigraph of $f$ are denoted by $G_f$ and Epi$_f$ respectively. The object $\mathcal{O}$ under observation is the spatial terrain surface described by $G_f$ in the world space $\mathcal{W} = \mathbb{R}^3$. The observation platform $\mathcal{P}$ on which the observers are attached corresponds to the *elevated surface of $G_f$* given by $G_{f_{h_v}}$, where $f_{h_v} \overset{\text{def}}{=} f + h_v$ with $h_v$ being a specified positive number. This implies that for any $x \in \Omega$, the observers are at a fixed vertical-height $h_v$ above the surface $G_f$.

From the extension of Proposition 3.1 to the case where $\dim(\Omega) = 2$, there exists a critical vertical-height $h_{vc}(x)$ for each $x \in \Omega$ such that total visibility is attainable. Consider the nontrivial case where $h_v < h_{vc}(x)$ for all $x \in \Omega$ so that the mobile observer must move to achieve total visibility. Let $I_{t_f} = [0, t_f]$ denote the observation time interval, where $t_f$ may be a finite fixed or variable terminal time. For simplicity, the mobile observer is represented by a point mass $M$. Its position in the world space $\mathcal{W} = \mathbb{R}^3$ at any time $t$ is specified by $p(t)$ whose representation with respect to a given orthonormal basis $B$ is denoted by $[x_1(t), x_2(t), x_3(t)]^T$, where $x_3(t)$ corresponds to the observer position along the $x_3$-axis at time $t$. The motion of the mobile observer can be described by Newton's law:

$$M\ddot{x}(t) + \nu_x \dot{x}(t) = u(t); \tag{5.1}$$

$$M\ddot{x}_3(t) + \nu_3(x(t), \dot{x}(t), x_3(t), \dot{x}_3(t)) = \xi(t) - Mg, \tag{5.2}$$

where $x(t) = [x_1(t), x_2(t)]^T$; $(u, \xi)$ is the external force with $u = [u_1, u_2]^T$ being the control; $-Mg$ is the gravitational force in the downward direction along the $x_3$-axis. $\nu_x$ is a given nonnegative friction coefficient; $\nu_3$ is a specified real-valued function of its arguments describing the $x_3$-component of the friction force. The variables with single and double overdots denote respectively their first and second derivatives with respect to time $t$. Assuming that the mobile observer is constrained to move on $G_f$ at all times without slipping, the mobile observer motion satisfies a holonomic constraint:

$$x_3(t) = f(x(t)) \text{ for all } t \in I_{t_f}, \tag{5.3}$$

and a state variable (position) constraint:

$$x(t) \in \Omega \text{ for all } t \in I_{t_f}. \tag{5.4}$$

Since $f$ is a $C_2$-function on $\Omega$, we may differentiate (5.3) twice with respect to $t$ to obtain

$$\dot{x}_3(t) = \nabla_x f(x(t))^T \dot{x}(t);$$
$$\ddot{x}_3(t) = \nabla_x f(x(t))^T \ddot{x}(t) + \dot{x}(t)^T H_f(x(t))\dot{x}(t), \tag{5.5}$$

where $\nabla_x$ denotes the gradient operator with respect to $x$, and $H_f(x(t))$ the Hessian matrix of $f$ with respect to $x$ evaluated at $x(t)$. Substituting (5.5) into (5.2) gives the required vertical component $\xi(t)$ of the external force for keeping the mobile observer on the surface $G_f$ at all times:

$$\xi(t) = M(\nabla_x f(x(t))^T \ddot{x}(t) + \dot{x}(t)^T H_f(x(t))\dot{x}(t))$$
$$+ \nu_3(x(t), \dot{x}(t), x_3(t), \nabla_x f(x(t))^T \dot{x}(t))/M + g). \tag{5.6}$$

Assuming that the mobile observer lies on $G_f$ at the starting time $t = 0$, then

$$x_3(0) = f(x(0)), \quad \dot{x}_3(0) = \nabla_x f(x(0))^T \dot{x}(0). \tag{5.7}$$

In the foregoing dynamic model of the mobile observer, we have assumed that the $x$-component of the friction force depend only on $\dot{x}$. In general, they may depend on both $(x, \dot{x})$ and $(x_3, \dot{x}_3)$. Also, to simplify the subsequent development, we have not considered the surface contact forces in the foregoing model.

Let $s_x(t) = (x(t), \dot{x}(t))$ denote the state of system (5.1) at time $t$. When necessary, $x_u(t; s_x(0))$ is used to indicate the dependence of a solution of (5.1) on the control $u$ and $s_x(0)$. Also, when it is unnecessary to indicate the dependence of a solution of (5.1) on $s_x(0)$, $x_u(t; s_x(0))$ is abbreviated by $x_u(t)$. A control $u = u(t)$ defined on a given time interval $I_{t_f}$ is said to be *admissible*, if it is a measurable function on $I_{t_f}$, and takes its values in the control region $U$, a nonempty compact subset of $\mathbb{R}^2$ containing the origin. Typical control regions are:

$$U_1 = \{(u_1, u_2) \in \mathbb{R}^2 : |u_i| \leq \bar{u}_i, i = 1, 2\};$$
$$U_2 = \{u \in \mathbb{R}^2 : \|u\| \leq \bar{M}\}, \tag{5.8}$$

where $\bar{M}, \bar{u}_i, i = 1, 2$ are given positive constants. The set of all admissible controls defined on $I_{t_f}$ is denoted by $\mathcal{U}_{\mathrm{ad}}(I_{t_f})$, where $t_f$ is a fixed or variable terminal time. Let the norm of any admissible control $u(\cdot)$ be defined by $\|u(\cdot)\| = \mathrm{ess.sup}_{t \in I_{t_f}} (|u_1(t)| + |u_2(t)|)$. Then, $\mathcal{U}_{\mathrm{ad}}(I_{t_f})$ is closed with respect to this norm. It can be deduced that for any $u(\cdot) \in \mathcal{U}_{\mathrm{ad}}(I_{t_f})$ and $s_x(0) = (x(0), \dot{x}(0)) \in \Omega \times \mathbb{R}^2$, (5.1) has a unique solution $x_u(t; s_x(0))$ which is twice differentiable with respect to $t$ almost everywhere on $I_{t_f}$. Moreover, $x_u(t; s_x(0))$ is continuous with respect to $u(\cdot)$. In what follows, we assume that no constraint is imposed on the magnitude of the vertical force $\xi$.

Let $K_x^+(t; s_x(0))$ denote the *attainable set of system* (5.1) *at time* $t > 0$ *starting from* $s_x(0) \in \mathbb{R}^4$ *at* $t = 0$ defined by

$$K_x^+(t; s_x(0)) = \{(x_u(t; s_x(0)), \dot{x}_u(t; s_x(0))) \in \mathbb{R}^4 : u(\cdot) \in \mathcal{U}_{\mathrm{ad}}(I_t)\}. \tag{5.9}$$

An admissible control $u(\cdot)$ defined on $I_{t_f}$ is said to be *extremal*, if $(x_u(t_f; s_x(0)), \dot{x}_u(t_f; s_x(0)))$ (the end-point of its corresponding response at time $t_f$) lies on $\partial K_x^+(t_f; s_x(0))$ (the boundary of $K_x^+(t_f; s_x(0))$). From the linearity of (5.1) and the assumption that $U$ is compact and convex, it can be readily shown that $K_x^+(t; s_x(0))$ is a compact convex subset of $\mathbb{R}^4$ for any $s_x(0) \in \mathbb{R}^4$ and any $t > 0$ with piecewise smooth boundary. Moreover, $K_x^+(t; s_x(0))$ varies continuously in $t$ with respect to the usual Euclidean metric $\rho_E$ on $\mathbb{R}$, and the Hausdorff metric $\rho_H$ on $2^{\mathbb{R}^4}$. It follows that $\Pi_{\mathbb{R}^2} K_x^+(t; s_x(0))$, the projection of $K_x^+(t; s_x(0))$ onto the configuration (position) space $\mathbb{R}^2$, is also compact and convex. Since $\Omega$ is generally non-convex, $\Pi_{\mathbb{R}^2} K_x^+(t; s_x(0)) \cap \Omega$ for any $t > 0$ is compact but may not be convex.

*Remark 5.1* The assumption that the control region $U$ is a compact (not necessarily convex) subset of $\mathbb{R}^4$ is sufficient to ensure the convexity of $K_x^+(t; s_x(0))$. This property can be proved by making use of Lyapunov's theorem on the range of a vector measure [1, p. 165]. Moreover, from the normality of system (5.1) [1], it can be shown that $K_x^+(t; s_x(0))$ is strictly convex.

### 5.1.1 Statement of Problems

Now, a few physically meaningful visibility-based optimal motion planning problems can be stated as follows:

**Problem 5.1 Minimum-Time Total Visibility Problem.** Let $\mathcal{U}_{\text{ad}} = \bigcup_{t_f \geq 0} \mathcal{U}_{\text{ad}}(I_{t_f})$ be the set of all admissible controls. Given $s_x(0) = (x(0), \dot{x}(0))$ or the initial state of the mobile observer with initial position $p(0) = (x(0), f(x(0))) \in G_f$ and initial velocity $v(0) = (\dot{x}(0), \nabla_x f(x(0))^T \dot{x}(0))$, find the smallest time $t_f^* \geq 0$ and an admissible control $u^* = u^*(t)$ defined on $I_{t_f^*}$ such that its corresponding motion or time-dependent path $\Gamma_{t_f^*} = \{(x_{u^*}(t), f(x_{u^*}(t))) \in \mathbb{R}^3 : t \in I_{t_f^*}\}$ on the surface $G_f$ satisfies the total visibility condition at $t_f^*$:

$$\bigcup_{t \in I_{t_f^*}} \mathcal{V}((x_{u^*}(t), f_{h_v}(x_{u^*}(t)))) = G_f \tag{5.10}$$

or alternatively,

$$\mu_2\left\{ \bigcup_{t \in I_{t_f^*}} \Pi_\Omega \mathcal{V}((x_{u^*}(t), f_{h_v}(x_{u^*}(t)))) \right\} = \mu_2\{\Omega\}, \tag{5.11}$$

where $\mu_2\{\sigma\}$ denotes the Lebesgue measure of the set $\sigma \subset \mathbb{R}^2$.

In the foregoing problem statement, condition (5.10) only involves the position $x_{u^*}(t)$, not the velocity $\dot{x}_{u^*}(t)$. In certain physical situations, it is required to move the mobile observer from one rest position to another, i.e. $\dot{x}_{u^*}(0) = 0$ and $\dot{x}_{u^*}(t_f^*) = 0$. Also, in planetary surface exploration, it is important to avoid paths with steep slopes. This suggests the inclusion of the following gradient constraint in Problem 5.1:

$$\|\nabla_x f(x_{u^*}(t))\| \leq f'_{\max} \text{ for all } t \in I_{t_f^*}, \tag{5.12}$$

where $f'_{\max}$ is a specified positive number. Now, if the set $\Omega' \overset{\text{def}}{=} \{x \in \Omega : \|\nabla_x f(x_{u^*}(t))\| \leq f'_{\max}\}$ is a simply connected compact subset of $\mathbb{R}^2$, then we may replace $\Omega$ in Problem 5.1 by $\Omega'$ to take care of constraint (5.12).

**Problem 5.2 Maximum Visibility Problem with Fixed Observation Time-Interval.** Given a finite observation time interval $I_{t_f}$ and $s_x(0) = (x(0), \dot{x}(0))$, or the initial state of the mobile observer with initial position $p(0) = (x(0), f(x(0))) \in G_f$ and initial velocity $v(0) = (\dot{x}(0), \nabla_x f(x(0))^T \dot{x}(0))$ at $t = 0$, find an admissible control $u^* = u^*(t)$ and its corresponding motion or time-dependent path $\Gamma_{t_f}^* = \{(x_{u^*}(t), f(x_{u^*}(t))) \in \mathbb{R}^3 : t \in I_{t_f}\} \subset G_f$ such that the visibility functional given by

$$J_1(u) = \int_0^{t_f} \mu_2\{\Pi_\Omega \mathcal{V}((x_u(t), f_{h_v}(x_u(t))))\}dt \qquad (5.13)$$

is defined, and satisfies $J_1(u^*) \geq J_1(u)$ for all $u(\cdot) \in \mathcal{U}_{\mathrm{ad}}(I_{t_f})$.
Another meaningful visibility functional is given by

$$J_2(u) = \mu_2\{ \bigcup_{t \in I_{t_f}} \Pi_\Omega \mathcal{V}((x_u(t), f_{h_v}(x_u(t))))\}. \qquad (5.14)$$

The foregoing problem with $J_1$ replaced by $J_2$ corresponds to selecting an admissible control $u^*(\cdot)$ such that the area of the union of the projected visibility sets on $\Omega$ for all the points along the corresponding time-dependent path $\Gamma^*_{t_f}$ is maximized.

## 5.1.2  Existence of Solutions

First, consider the Minimum-time Total Visibility Problem 5.1. The following result provides a sufficient condition under which the set of all motions or time-dependent paths $\Gamma_{t_f}$ of (5.1) generated by admissible controls $u(\cdot)$ satisfying the total visibility condition (5.10) is nonempty.

**Proposition 5.1** *Given an initial state $s_x(0) = (x(0), \dot{x}(0))$ of (5.1) such that its corresponding position $x(0) \in \Omega$, if*

$$\bigcup_{t \in I_{t_f}} \Pi_\Omega K_x^+(t; s_x(0)) = \Omega \quad \text{for some finite } t_f > 0, \qquad (5.15)$$

*then there exists a smooth time-dependent path $\tilde{\Gamma}_{t_f} = \{(\tilde{x}(t), \tilde{x}_3(t)) \in \mathbb{R}^3 : \tilde{x}_3(t) = f(\tilde{x}(t)), t \in I_{t_f}\} \subset G_f$. This path satisfies the total visibility condition (5.10) provided that*

$$\tilde{u}(t) \overset{\text{def}}{=} M\ddot{\tilde{x}}(t) + \nu_x \dot{\tilde{x}}(t) \in U \quad \text{for almost all } t \in I_{t_f}. \qquad (5.16)$$

*Proof* From Theorem 3.3, there exists a finite point set $P_N = \{x^{(i)}, i = 1, \dots, N\} \subset \Omega$ such that $G_f$ is totally visible from the finite set of observation points $\{(x^{(1)}, f_{h_v}(x^{(1)})), \dots, (x^{(N)}, f_{h_v}(x^{(N)}))\}$. If the initial point $x(0) \in P_N$, we set $x^{(1)} = x(0)$. Under the assumption that $\Omega$ is simply connected, it is always possible to construct a smooth Jordan arc $\mathcal{A} = \{\tilde{x}(\lambda) \in \Omega : 0 \leq \lambda \leq 1\}$ (e.g. an arc represented by spline functions) passing through all the points in $P_N$. If $x(0) \notin P_N$, we augment $P_N$ by this point, and proceed with the construction of a smooth Jordan arc $\mathcal{A}$ starting from $x(0)$ and passing through all the points in $P_N$. Now, let $t = \lambda t_f, 0 \leq \lambda \leq 1$, where $t_f$ is the smallest terminal time such that $\bigcup_{t \in I_{t_f}} \Pi_\Omega K_x^+(t; s_x(0)) \subset \Omega$. If condition (5.15) is satisfied, then $\tilde{u}(\cdot)$ is an admissible control that generates the time-dependent path $\tilde{\Gamma}_{t_f} = \{(\tilde{x}(t), f_{h_v}(\tilde{x}(t))) \in \mathbb{R}^3 : t \in I_{t_f}\} \subset G_f$ satisfying the total visibility condition (5.10). □

**Remark 5.2** In the proof of Proposition 5.1, linear scaling between $t$ and $\lambda$ is used. The satisfaction of condition (5.15) by the constructed Jordan arc $\mathcal{A}$ generally depends on the ordering of the points in $P_N$ and the time-scaling. Therefore, it is desirable to consider all permutations of the points in $P_N$, and other forms of scaling between $t$ and $\lambda$.

**Remark 5.3** Condition (5.15) in Proposition 5.1 is not necessary for the existence of a smooth time-dependent path $\tilde{\Gamma}_{t_f}$ that satisfies both conditions (5.10) and (5.15). In fact, such a smooth path may exist when $\bigcup_{t \in I_{t_f}} \Pi_\Omega K_x^+(t; s_x(0)) \subset \Omega$. Obviously, there always exists a smooth time-dependent path $\tilde{\Gamma}_{t_f}$ that satisfies condition (5.15) but not the total visibility condition (5.10). A trivial example is the case where the control region $U = \{0\} \subset \mathbb{R}^2$, and $s_x(0) = (x(0), 0)$ so that $\Pi_\Omega K_x^+ (t; s_x(0)) = \{x(0)\} \subset \Omega$ for all $t \geq 0$. Let $\mathcal{P}$ denote the set of all motions or time-dependent paths $\Gamma_{t_f} \subset G_f$ starting from $(x(0), f_{h_v}(x(0)))$ at $t = 0$, generated by admissible controls $u(\cdot) \in \mathcal{U}_{ad}(I_{t_f})$. Let $\mathcal{P}_{TV}$ denote the subset of time-dependent paths in $\mathcal{P}$ satisfying the total visibility condition (5.10). Then a necessary condition for the existence of a solution to Problem 5.1 is that $\mathcal{P}_{TV}$ is nonempty.

**Theorem 5.1** *Suppose there exists an admissible control $\tilde{u}(\cdot)$ for (5.1) defined on some finite time interval $I_{\tilde{t}_f}$ with corresponding motion or time-dependent path*

$$\tilde{\Gamma}_{\tilde{t}_f} = \{(x_{\tilde{u}}(t), x_{\tilde{u},3}(t)) \in \mathbb{R}^3 : x_{\tilde{u},3}(t) = f(x_{\tilde{u}}(t)), t \in I_{\tilde{t}_f}\} \qquad (5.17)$$

*such that $\bigcup_{t \in I_{\tilde{t}_f}} \mathcal{V}((x_{\tilde{u}}(t), f_{h_v}(x_{\tilde{u}}(t)))) = G_f$. If the mapping $x \to \mathcal{V}((x, f_{h_v}(x)))$ on $\Omega$ into $2^{G_f}$ is continuous with respect to metrics $\rho_E$ and $\rho_H$. Then, there exists a minimum time $t_f^* \leq \tilde{t}_f$ and an optimal control $u^*(\cdot) \in \mathcal{U}_{ad}(I_{t_f^*})$ such that*

$$\bigcup_{t \in I_{t_f^*}} \mathcal{V}((x_{u^*}(t), f_{h_v}(x_{u^*}(t)))) = G_f. \qquad (5.18)$$

*Proof* Let $t_f^* = \inf\{t_f \geq 0 : \bigcup_{t \in I_{t_f}} \mathcal{V}((x_u(t), f_{h_v}(x_u(t)))) = G_f, u(\cdot) \in \mathcal{U}_{ad}(I_{t_f})\}$. From the assumption that there exists an admissible control $\tilde{u}(\cdot)$ defined on $I_{\tilde{t}_f}$ such that

$$\bigcup_{t \in I_{\tilde{t}_f}} \mathcal{V}((x_{\tilde{u}}(t), f_{h_v}(x_{\tilde{u}}(t)))) = G_f$$

for some finite $\tilde{t}_f \geq 0$, we have $0 \leq t_f^* \leq \tilde{t}_f$. Thus, there exists a non-increasing sequence $\{t_f^{(k)}\}$ converging to $t_f^*$ and a sequence of admissible controls $\{u^{(k)}(\cdot)\}$ such that

$$\bigcup_{t \in I_{t_f}^{(k)}} \mathcal{V}(x_{u^{(k)}}(t), f_{h_v}(x_{u^{(k)}}(t)))) = G_f$$

for all $k = 1, 2, \ldots$. Since $x_u(t)$ is continuous with respect to $u(\cdot) \in \mathcal{U}_{\mathrm{ad}}(I_t)$ (a closed set), and the mapping $x \to \mathcal{V}(x, f_{h_v}(x)))$ on $\Omega$ into $2^{G_f}$ is assumed to be continuous with respect to metrics $\rho_E$ and $\rho_H$, it follows that

$$\lim_{k \to \infty} \bigcup_{t \in I_{t_f}^{(k)}} \mathcal{V}((x_{u^{(k)}}(t), f_{h_v}(x_{u^{(k)}}(t)))) = \bigcup_{t \in I_{t_f}^*} \mathcal{V}((x_{u^*}(t), f_{h_v}(x_{u^*}(t))))$$

exists and is equal to $G_f$ for some $u^*(\cdot) \in \mathcal{U}_{\mathrm{ad}}(I_{t_f^*})$, where the convergence is defined with respect to the Hausdorff metric $\rho_H$. By definition of $t_f^*$, $u^*(\cdot)$ is optimal.    □

Now, consider the Fixed Observation Time-interval Maximum Visibility Problem 5.2 with visibility functional $J_1$. Under the assumption that the observation platform's vertical height $h_v < h_{vc}(x)$ (the critical vertical-height at $x$) for all $x \in \Omega$, $\mu_2\{\Pi_\Omega \mathcal{V}((x_u(t), f_{h_v}(x_u(t)))\} < \mu_2(\Omega)$ for all $t \in I_{t_f}$. Thus, $0 \le J_1(u) < t_f \mu_2(\Omega)$. Since the set-valued mapping $x \to \mathcal{V}(x, f_{h_v}(x))$ on $\Omega$ into $2^{G_f}$ may be discontinuous with respect to metrics $\rho_E$ and $\rho_H$, hence the mapping $t \to \Pi_\Omega \mathcal{V}((x_u(t), f_{h_v}(x_u(t))))$ on $I_{t_f}$ into $2^\Omega$ may be discontinuous also. It follows that the mapping $t \to \mu_2\{\Pi_\Omega \mathcal{V}((x_u(t), f_{h_v}(x_u(t))))\}$ on $I_{t_f}$ into $\mathbb{R}^+ \overset{\mathrm{def}}{=} [0, +\infty[$ may be discontinuous for some $u(\cdot) \in \mathcal{U}_{\mathrm{ad}}(I_{t_f})$. Nevertheless $J_1$ is maximized if there exists an admissible $u(\cdot)$ defined on $I_{t_f}$ that maximizes $\mu_2\{\Pi_\Omega \mathcal{V}((x_u(t), f_{h_v}(x_u(t))))\}$ point-wise for almost all $t \in I_{t_f}$. Now, consider the compact set $V_\Omega(t) \overset{\mathrm{def}}{=} \Pi_\Omega(\partial K_x^+(t; s_x(0))$ for $t \in I_{t_f}$. Although the mapping $x \to \mu_2\{\Pi_\Omega \mathcal{V}((x, f_{h_v}(x)))\}$ with domain restricted to $V_\Omega(t)$ is generally discontinuous, the set $\hat{V}_\Omega(t) \overset{\mathrm{def}}{=} \{x \in V_\Omega(t) : \mu_2\{\Pi_\Omega \mathcal{V}((x, f_{h_v}(x)))\} = \hat{\mu}_2(t)\}$ is nonempty, where

$$\hat{\mu}_2(t) \overset{\mathrm{def}}{=} \mathrm{ess.\ sup}\{\mu_2\{\Pi_\Omega \mathcal{V}((x, f_{h_v}(x)))\} : x \in V_\Omega(t)\}. \tag{5.19}$$

Now, if a time-dependent path $\hat{\Lambda} = \{\hat{x}(t) \in \hat{V}_\Omega(t) : t \in I_{t_f}\}$ can be constructed such that

$$\hat{u}(t) \overset{\mathrm{def}}{=} M\ddot{\hat{x}}(t) + \nu_x \dot{\hat{x}}(t) \in U \quad \text{for almost all } t \in I_{t_f} \tag{5.20}$$

is satisfied, then $\hat{u}(\cdot)$ is an admissible control that maximizes $J_1$ implying that $\hat{u}(\cdot)$ is an optimal control. The foregoing construction gives a sufficient condition for the existence of a solution for Problem 5.2 with visibility functional $J_1$. The result can be summarized as a theorem.

**Theorem 5.2** *If there exists a motion or time-dependent path $\hat{\Lambda} = \{\hat{x}(t) \in \hat{V}_\Omega(t) : t \in I_{t_f}\}$ such that the corresponding control $\hat{u}(\cdot)$ satisfies (5.20), then Problem 5.2 with visibility $J_1$ has a solution.*

A similar result can be established for Problem 5.2 with visibility functional $J_2$. The underlying idea is useful in the computation of solutions for Problem 5.2 to be discussed later.

### 5.1.3 Optimality Conditions

First, necessary conditions for optimality associated with the Minimum-time Total Visibility Problem 5.1 will be derived under the assumption that a solution exists. Consider the quantity:

$$w_u(t; s_x(0)) \overset{\text{def}}{=} \mu_2 \{ \bigcup_{\tau \in I_t} \Pi_\Omega \mathcal{V}((x_u(\tau; s_x(0)), f_{h_v}(x_u(\tau; s_x(0))))) \} \qquad (5.21)$$

with $w_u(0; s_x(0)) = \mu_2\{\Pi_\Omega \mathcal{V}((x(0), f_{h_v}(x(0))))\}$. Evidently, Problem 5.1 is equivalent to finding the smallest time $t_f^* > 0$ and an admissible control $u^* = u^*(t)$ defined on $I_{t_f^*}$ such that $w_{u^*}(t_f^*; s_x(0)) = \mu_2\{\Omega\}$. Given $s_x(0)$ and an admissible control $u(\cdot)$ defined on $I_{t_f}$ such that $x_u(t; s_x(0)) \in \Omega$ for all $t \in I_{t_f}$, $w_u = w_u(t; s_x(0))$ is a monotonically increasing but not necessarily continuous function of $t$. Let $K_w^+(t; s_x(0))$ denote the attainable set corresponding to $w$ at time $t$ defined by

$$K_w^+(t; s_x(0)) \overset{\text{def}}{=} \{w_u(t; s_x(0)) \in \mathbb{R}^+ : x_u(\tau; s_x(0)) \in \Omega \text{ for all } \tau \in I_t, u(\cdot) \in \mathcal{U}_{\text{ad}}(I_t)\} \qquad (5.22)$$

with $K_w^+(0; s_x(0))$ being the singleton $\{\mu_2\{\Pi_\Omega \mathcal{V}((x(0), f_{h_v}(x(0))))\}\}$. In general, $K_w^+(t; s_x(0))$ may be the union of disjoint, bounded, but not necessarily closed subintervals of $\mathbb{R}^+$. Moreover, the mapping $t \to \overline{K_w^+(t; s_x(0))}$ (the closure of $K_w^+(t; s_x(0))$) may be discontinuous with respect to the usual Euclidean metric $\rho_E$ on $\mathbb{R}$ and the Hausdorff metric $\rho_H$ on $2^{\mathbb{R}^+}$.

**Proposition 5.2** *Assume that an optimal control $u^*(\cdot)$ defined on $I_{t_f^*}$ for Problem 5.1 exists. Then, $w_{u^*}(t_f^*; s_x(0))$ is the upper boundary point of $co(K_w^+(t_f^*; s_x(0)))$, where $co(S)$ denotes the convex hull of the set $S$.*

*Proof* From the definition of $w_u(t; s_x(0))$ given by (5.21), we have

$$0 < \mu_2\{\Pi_\Omega \mathcal{V}((x(0), f_{h_v}(x(0))))\} \leq w_u(t_f^*; s_x(0)) \leq \mu_2(\Omega) \qquad (5.23)$$

for all $u(\cdot) \in \mathcal{U}_{ad}(I_{t_f^*})$ such that $x_u(t; s_x(0)) \in \Omega$ for all $t \in I_{t_f^*}$. Suppose that $w_u(t_f^*; s_x(0))$ is an interior point of $co(K_w^+(t_f^*; s_x(0)))$ satisfying $w_u(t_f^*; s_x(0)) = \mu_2(\Omega)$. Then there exist a $\delta$-neighborhood $\mathcal{N}_\delta(w_{u^*}(t_f^*; s_x(0)))$ of $w_{u^*}(t_f^*; s_x(0))$ such that

$$\mathcal{N}_\delta(w_{u^*}(t_f^*; s_x(0))) \subset co(K_w^+(t_f^*; s_x(0))),$$

and a point $w_u(t_f^*; s_x(0)) \in \mathcal{N}_\delta(w_{u^*}(t_f^*; s_x(0)))$ satisfying $w_u(t_f^*; s_x(0)) > \mu_2(\Omega)$, which contradicts the upper bound in (5.23).                                                $\square$

Another necessary condition for optimality can be derived by considering the augmented system:

$$\frac{d}{dt}\begin{bmatrix} x \\ \dot{x} \\ w \end{bmatrix} = \begin{bmatrix} \dot{x} \\ (-\nu_x\dot{x} + u)/M \\ g(x) \end{bmatrix}, \tag{5.24}$$

with initial state at $t = 0$ given by

$$s_{(x,w)}(0) \overset{\text{def}}{=} (x(0), \dot{x}(0), \mu_2\{\Pi_\Omega \mathcal{V}((x(0), f_{h_v}(x(0))))\}) \in \Omega \times \mathbb{R}^2 \times \mathbb{R}^+, \tag{5.25}$$

where

$$g(x_u(t)) \overset{\text{def}}{=} \lim_{\delta t \to 0^+} \sup \frac{1}{\delta t}[w_u(t + \delta t; s_x(0)) - w_u(t; s_x(0))]$$

$$= \lim_{\delta t \to 0^+} \sup \frac{1}{\delta t}\Bigg[\mu_2\Bigg\{ \bigcup_{\tau \in I_{t+\delta t}} \Pi_\Omega \mathcal{V}((x_u(\tau), f_{h_v}(x_u(\tau)))) \Bigg\}$$

$$- \mu_2\Bigg\{ \bigcup_{\tau \in I_t} \Pi_\Omega \mathcal{V}((x_u(\tau), f_{h_v}(x_u(\tau)))) \Bigg\}\Bigg]. \tag{5.26}$$

From the elementary identity for arbitrary connected subsets $S_1$ and $S_2$ in Lemma 4.1, we set

$$S_1 = \bigcup_{\tau \in I_{t+\delta t}} \Pi_\Omega \mathcal{V}((x_u(\tau), f_{h_v}(x_u(\tau)))), \quad S_2 = \bigcup_{\tau \in I_t} \Pi_\Omega \mathcal{V}((x_u(\tau), f_{h_v}(x_u(\tau)))))$$

and obtain

$$\mu_2\Bigg\{ \bigcup_{\tau \in I_{t+\delta t}} \Pi_\Omega \mathcal{V}((x_u(\tau), f_{h_v}(x_u(\tau)))) \Bigg\} - \mu_2\Bigg\{ \bigcup_{\tau \in I_t} \Pi_\Omega \mathcal{V}((x_u(\tau), f_{h_v}(x_u(\tau)))) \Bigg\}$$

$$= \mu_2\Bigg\{ [\bigcup_{\tau \in I_{t+\delta t}} \Pi_\Omega \mathcal{V}((x_u(\tau), f_{h_v}(x_u(\tau))))]^c \cap [\bigcup_{\tau \in I_t} \Pi_\Omega \mathcal{V}((x_u(\tau), f_{h_v}(x_u(\tau))))] \Bigg\}$$

$$- \mu_2\Bigg\{ [\bigcup_{\tau \in I_{t+\delta t}} \Pi_\Omega \mathcal{V}((x_u(\tau), f_{h_v}(x_u(\tau))))] \cap [\bigcup_{\tau \in I_t} \Pi_\Omega \mathcal{V}((x_u(\tau), f_{h_v}(x_u(\tau))))]^c \Bigg\}.$$

$$\tag{5.27}$$

It follows that $g(x_u(t))$ can be rewritten as

$$
\begin{aligned}
g(x_u(t)) = \lim_{\delta t \to 0^+} \sup \frac{1}{\delta t} \Bigg[ &\mu_2 \Bigg\{ \Big[ \bigcup_{\tau \in I_{t+\delta t}} \Pi_\Omega \mathcal{V}((x_u(\tau), f_{h_v}(x_u(\tau)))) \Big]^c \\
&\cap \Big[ \bigcup_{\tau \in I_t} \Pi_\Omega \mathcal{V}((x_u(\tau), f_{h_v}(x_u(\tau)))) \Big] \Bigg\} \\
- \mu_2 \Bigg\{ &\Big[ \bigcup_{\tau \in I_{t+\delta t}} \Pi_\Omega \mathcal{V}((x_u(\tau), f_{h_v}(x_u(\tau)))) \Big] \\
&\cap \Big[ \bigcup_{\tau \in I_t} \Pi_\Omega \mathcal{V}((x_u(\tau), f_{h_v}(x_u(\tau)))) \Big]^c \Bigg\} \Bigg].
\end{aligned}
\tag{5.28}
$$

The target set is specified by $\mathcal{T} = \{(x, \dot{x}, w) \in \Omega \times \mathbb{R}^2 \times \mathbb{R}^+ : w = \mu_2\{\Omega\}\}$. Thus, Problem 5.1 can be restated in the form of a standard time-optimal control problem, i.e. find an admissible control $u^*(\cdot)$ which steers the initial state $s_{(x,w)}(0)$ of system (5.24) at $t = 0$ to the target set $\mathcal{T}$ in minimum time $t_f^*$ subject to the constraint that $x_{u^*}(t; s_x(0)) \in \Omega$ for all $t \in I_{t_f^*}$.

Let $K_{(x,w)}^+(t; s_{(x,w)}(0))$ denote the attainable set of (5.24) at time $t$ starting from $s_{(x,w)}(0)$ at $t = 0$ defined by $\{(x_u(t; s_x(0)), \dot{x}_u(t; s_x(0)), w_u(t; s_x(0))) \in \mathbb{R}^5 : u(\cdot) \in \mathcal{U}_{\text{ad}}(I_{t_f})\}$. A necessary condition for which an optimal control associated with Problem 5.1 must satisfy can be obtained under the assumption that $w_u = w_u(t; s_x(0))$ is a continuous function of $t$.

**Theorem 5.3** *If $w_u = w_u(t; s_x(0))$ is continuous in $t$ for any admissible control $u(\cdot)$ such that $x_u(t; s_x(0)) \in \Omega$ for all $t \in I_{t_f}$, where $t_f$ depends on $u(\cdot)$, then the time-optimal control $u^*$ for Problem 5.1 is an extremal control that steers $s_{(x,w)}(0)$ to the boundary of the attainable set of (5.24) at time $t_f^*$, i.e. $(x, \dot{x}, w)(t_f^*) \in \partial K_{(x,w)}^+(t_f^*; s_{(x,w)}(0))$.*

*Proof* Under the assumption on the continuity of $w_u = w_u(t; s_x(0))$, $K_{(x,w)}^+(t; s_{(x,w)}(0))$ is a compact subset of $\mathbb{R}^4 \times \mathbb{R}^+$ that varies continuously with $t$. The solution to Problem 5.1 is obtained when $\mathcal{T} \cap K_{(x,w)}^+(t; s_{(x,w)}(0))$ is nonempty for the first or smallest $t = t_f^* > 0$. Let $(x, \dot{x}, w)(t_f^*)$ be a solution of (5.24) at $t_f^*$ corresponding to the admissible control $u^*(\cdot)$ defined on $I_{t_f^*}$. Then, $(x, \dot{x}, w)(t_f^*)$ must be a boundary point of $K_{(x,w)}^+(t_f^*; s_{(x,w)}(0))$. Otherwise, from the continuity of $K_{(x,w)}^+(t; s_{(x,w)}(0))$ with respect to $t$, there exists a $t_f < t_f^*$ such that $\mathcal{T} \cap K_{(x,w)}^+(t_f; s_{(x,w)}(0))$ is nonempty, which contradicts the definition of $t_f^*$. □

*Remark 5.4* We observe that the attainable sets of (5.1) are related to those of the augmented system (5.24) by

$$
K_x^+(t; s_x(0)) = \Pi_{\mathbb{R}^4} K_{(x,w)}^+(t; s_{(x,w)}(0)).
\tag{5.29}
$$

Evidently, a boundary point of $K_x^+(t; s_x(0))$ corresponds to some boundary point of $K_{(x,w)}^+(t; s_{(x,w)}(0))$. But a boundary point of $K_{(x,w)}^+(t; s_{(x,w)}(0))$ may correspond to an interior point of $K_x^+(t; s_x(0))$. Thus, an extremal control of the augmented system (5.24) may not be an extremal control of (5.1). Hence, an optimal control for Problem 5.1 may not be an extremal control of (5.1).

Now, we define the Hamiltonian associated with the augmented system (5.24) by:

$$\mathcal{H}(x, \dot{x}, w, \eta, u) = \eta^T (\dot{x}, (-\nu_x \dot{x} + u)/M, g(x))^T, \tag{5.30}$$

where $\eta = [\eta_1, \ldots, \eta_5]^T$ corresponds to the state of the adjoint system:

$$\dot{\eta} = -\nabla_{(x,\dot{x},w)} \mathcal{H}, \tag{5.31}$$

where $\nabla_{(x,\dot{x},w)}$ denotes the gradient operator with respect to $(x, \dot{x}, w)$.

If the real-valued function $x \to g(x)$ on $\Omega \to \mathbb{R}^+$ is smooth, then the following necessary condition for optimality follows from the Pontryagin Maximum Principle [1–3]:

**Theorem 5.4** *Suppose that the function $x \to g(x)$ on $\Omega \to \mathbb{R}^+$ is $C_1$. Let $u^* = u^*(t)$ be an optimal control for Problem 5.1 with corresponding response $x^* = x^*(t)$. Then there exists an absolutely continuous function $\eta^* = \eta^*(t)$ satisfying (5.31) given explicitly by*

$$\frac{d}{dt} \begin{bmatrix} \eta_1 \\ \eta_2 \\ \eta_3 \\ \eta_4 \\ \eta_5 \end{bmatrix} = - \begin{bmatrix} \eta_5 \partial g/\partial x_1 \\ \eta_5 \partial g/\partial x_2 \\ \eta_1 - (\nu_x/M)\eta_3 \\ \eta_2 - (\nu_x/M)\eta_4 \\ 0 \end{bmatrix} \tag{5.32}$$

*for almost all $t \in I_{t_f}$ with*

$$\mathcal{H}(x^*(t), \dot{x}^*(t), w^*(t), \eta^*(t), u^*(t))$$
$$= \mathcal{M}(x^*(t), \dot{x}^*(t), w^*(t), \eta^*(t)) \overset{\text{def}}{=} \max_{u \in U} \mathcal{H}(x^*(t), \dot{x}^*(t), w^*(t), \eta^*(t), u) \tag{5.33}$$

*for almost all $t \in I_{t_f}$. Moreover,*

$$\mathcal{M}(x^*(t), \dot{x}^*(t), w^*(t), \eta^*(t)) \equiv 0 \text{ on } I_{t_f}. \tag{5.34}$$

In view of Theorem 5.4 and (5.30), the optimal control $u^*$ is a function of $\eta_3^*$ and $\eta_4^*$ only. For the special case where the control region $U = U_1$, $u^*(t)$ takes on the form:

$$u_i^*(t) = \bar{u}_i \text{sgn}(\eta_{i+2}^*(t)), \quad i = 1, 2. \tag{5.35}$$

Equations (5.24), (5.32)–(5.35) with terminal condition $w(t_f) = \mu_2\{\Omega\}$ and initial condition $(x, \dot{x}, w)\,(0) = (x(0), \dot{x}(0), \mu_2\{\Pi_\Omega \mathcal{V}((x(0), f_{h_v}(x(0)))\})$ constitute a family of two-point-boundary-value problems (TPBVP) with the terminal time $t_f$ as a variable parameter. The optimal trajectory $(x^*, \dot{x}^*, w^*)(\cdot)$ is a solution of the TPBVP with the smallest terminal time $t_f^*$ which is determined by the transversality condition: $-\eta(t_f^*) = (0, \kappa)^T$ for some $\kappa > 0$ (the normal to the target set $\mathcal{T}$ at $(x^*, \dot{x}^*, w^*)(t_f^*)$ in the positive $w$-direction), or

$$\eta^*(t_f^*) = 0, \quad \eta_{n+1}^*(t_f^*) = -\kappa. \tag{5.36}$$

Now, consider the Fixed Observation Time-interval Maximum Visibility Problem 5.2 with visibility functional $J_1$. Assume that an optimal control $u^* = u^*(t)$ defined on $I_{t_f}$ exists. Let $\delta u$ be a control perturbation such that $u = u^* + \delta u$ is admissible. Let the solutions of (5.1) at time $t$ corresponding to $u$ and $u^*$, and the same initial state $s_x(0)$ be denoted by $x_u(t)$ and $x_{u^*}(t)$ respectively.

To derive optimality conditions for this problem, consider

$$\Delta J_1 \overset{\text{def}}{=} J_1(u^*) - J_1(u^* + \delta u) = \int_0^{t_f} (\mu_2\{\Pi_\Omega \mathcal{V}(p_{u^*}(t))\} - \mu_2\{\Pi_\Omega \mathcal{V}(p_{u^*+\delta u}(t))\})dt, \tag{5.37}$$

where $p_u(t) = (x_u(t), f_{h_v}(x_u(t)))$. Using the set identities in Lemma 4.1, we obtain

$$\begin{aligned}
\mu_2\{\Pi_\Omega \mathcal{V}(p_{u^*}(t))\} &- \mu_2\{\Pi_\Omega \mathcal{V}(p_{u^*+\delta u}(t))\} \\
&= \mu_2\{(\Pi_\Omega \mathcal{V}(p_{u^*+\delta u}(t)))^c \cap \Pi_\Omega \mathcal{V}(p_{u^*}(t))\} \\
&\quad - \mu_2\{(\Pi_\Omega \mathcal{V}(p_{u^*}(t)))^c \cap \Pi_\Omega \mathcal{V}(p_{u^*+\delta u}(t))\},
\end{aligned} \tag{5.38}$$

$\Delta J_1$ can be rewritten as:

$$\begin{aligned}
\Delta J_1 = \int_0^{t_f} (&\mu_2\{(\Pi_\Omega \mathcal{V}(p_{u^*+\delta u}(t)))^c \cap \Pi_\Omega \mathcal{V}(p_{u^*}(t))\} \\
&- \mu_2\{(\Pi_\Omega \mathcal{V}(p_{u^*}(t)))^c \cap \Pi_\Omega \mathcal{V}(p_{u^*+\delta u}(t))\})dt.
\end{aligned} \tag{5.39}$$

Thus, a sufficient but not necessary condition for optimality is given by

$$\mu_2\{(\Pi_\Omega \mathcal{V}(p_{u^*+\delta u}(t)))^c \cap \Pi_\Omega \mathcal{V}(p_{u^*}(t))\} \geq \mu_2\{(\Pi_\Omega \mathcal{V}(p_{u^*}(t)))^c \cap \Pi_\Omega \mathcal{V}(p_{u^*+\delta u}(t))\} \tag{5.40}$$

for almost all $t \in I_{t_f}$ and all admissible $u^* + \delta u$, where

$$p_{u^*+\delta u}(t) = (x_{u^*+\delta u}(t), f_{h_v}(x_{u^*+\delta u}(t))). \tag{5.41}$$

The perturbed position $x_{u^*+\delta u}(t)$ can be written in the form:

$$x_{u^*+\delta u}(t) = x_{u^*}(t) + \delta x(t), \tag{5.42}$$

with

$$x_{u^*}(t) = \Phi(t)s_x(0) + \frac{1}{M}\int_0^t \Phi(t-\tau)Bu^*(\tau)d\tau,$$

$$\delta x(t) = \frac{1}{M}\int_0^t \Phi(t-\tau)B\delta u(\tau)d\tau, \qquad\qquad (5.43)$$

where $s_x(0) = [x_1(0), \dot{x}_1(0), x_2(0), \dot{x}_2(0)]^T$, $u^* = [u_1^*, u_2^*]^T$, $\delta u = [\delta u_1, \delta u_2]^T$;

$$\Phi(t) = \begin{cases} \begin{bmatrix} 1 & \frac{M}{\nu_x}(1-\exp(-\frac{\nu_x t}{M})) & 0 & 0 \\ 0 & 0 & 1 & \frac{M}{\nu_x}(1-\exp(-\frac{\nu_x t}{M})) \end{bmatrix}, & \text{for } \nu_x > 0; \\[2em] \begin{bmatrix} 1 & t & 0 & 0 \\ 0 & 0 & 1 & t \end{bmatrix}, & \text{for } \nu_x = 0, \end{cases} \qquad (5.44)$$

and

$$B = \begin{bmatrix} 0 & 0 \\ 1 & 0 \\ 0 & 0 \\ 0 & 1 \end{bmatrix}. \qquad\qquad (5.45)$$

Now, consider perturbed admissible controls of the form $u^* + \alpha\delta u$, where $\delta u$ is a given control perturbation, and $0 \le \alpha < 1$. In view of (5.41)–(5.45), the corresponding perturbed position can be written in the form:

$$p_{u^*+\alpha\delta u}(t) \overset{\text{def}}{=} (x_{u^*}(t) + \alpha\delta x(t), f_{h_v}(x_{u^*}(t) + \alpha\delta x(t)))$$

$$= (x_{u^*}(t), f_{h_v}(x_{u^*}(t)) + \alpha(\delta x(t), \nabla_x f_{h_v}(x_{u^*}(t))^T \delta x(t)) + o(\alpha). \qquad (5.46)$$

If the real-valued function $x \to \mu_2\{\Pi_\Omega \mathcal{V}((x, f_{h_v}(x)))\}$ on $\Omega \to \mathbb{R}^+$ is $C_1$, then $G_{\mu_2}$ (graph of the mapping $\mu_2$) is a two-dimensional $C_1$-surface in $\mathbb{R}^3$. It follows that the Gateaux differential of $J_1$ at $x_{u^*}(\cdot)$ with increment $\delta x(\cdot)$ exists. Thus, we have the following necessary condition for optimality:

**Theorem 5.5** *Suppose that an optimal control $u^* = u^*(t)$ defined on $I_{t_f}$ for Problem 5.2 with visibility functional $J_1$ exists, and the function $x \to \mu_2\{\Pi_\Omega \mathcal{V}((x, f_{h_v}(x)))\}$ on $\Omega \to \mathbb{R}^+$ is $C_1$. Then $u^*(\cdot)$ must satisfy the following variational inequality:*

$$DJ_1(u^*; \delta u)$$

$$= \int_0^{t_f} \lim_{\alpha \to 0^+} \frac{1}{\alpha}(\mu_2\{S(p_{u^*+\alpha\delta u}(t))^c \cap S(p_{u^*}(t))\}$$

$$- \mu_2\{S(p_{u^*}(t))^c \cap S(p_{u^*+\alpha\delta u}(t))\})dt \ge 0 \qquad (5.47)$$

*for all admissible $u^* + \alpha\delta u$, $0 \le \alpha < 1$, where $S(p) = \Pi_\Omega \mathcal{V}(p)$.*

Another necessary condition for optimality can be obtained from classical optimal control theory [1] by introducing a new state variable $y$. The evolution of $y(t)$ with time $t$ is described by

$$\dot{y}(t) = \mu_2\{\Pi_\Omega \mathcal{V}((x_u(t), f_{h_v}(x_u(t))))\}, \quad y(0) = 0. \tag{5.48}$$

Thus, $J_1(u) = y(t_f)$. Let the Hamiltonian associated with the augmented system (5.1) and (5.48) be defined by

$$\mathcal{H}(x, \dot{x}, \eta, u) = \mu_2\{\Pi_\Omega \mathcal{V}((x, f_{h_v}(x)))\} + \eta^T(\dot{x}, (-\nu_x \dot{x} + u)/M)^T, \tag{5.49}$$

where $\eta = [\eta_1, \ldots, \eta_4]^T$ corresponds to the state of the adjoint system:

$$\dot{\eta} = -\nabla_{(x, \dot{x})}\mathcal{H}. \tag{5.50}$$

Again, if the function $x \to \mu_2\{\Pi_\Omega \mathcal{V}((x, f_{h_v}(x)))\}$ on $\Omega \to \mathbb{R}^+$ is smooth, then we have the following necessary condition for optimality:

**Theorem 5.6** *Suppose that the function $x \to \mu_2\{\Pi_\Omega \mathcal{V}((x, f_{h_v}(x)))\}$ on $\Omega \to \mathbb{R}^+$ is $C_1$. Let $u^* = u^*(t)$ be an optimal control for Problem 5.2 with visibility functional $J_1$ and corresponding response $x^* = x^*(t)$. Then there exists an absolutely continuous function $\eta^* = \eta^*(t)$ satisfying (5.50) given explicitly by*

$$\frac{d}{dt}\begin{bmatrix} \eta_1 \\ \eta_2 \\ \eta_3 \\ \eta_4 \end{bmatrix} = -\begin{bmatrix} \partial\mu_2\{\Pi_\Omega \mathcal{V}((x, f_{h_v}(x)))\}/\partial x_1 \\ \partial\mu_2\{\Pi_\Omega \mathcal{V}((x, f_{h_v}(x)))\}/\partial x_2 \\ \eta_1 - (\nu_x/M)\eta_3 \\ \eta_2 - (\nu_x/M)\eta_4 \end{bmatrix} \tag{5.51}$$

*for almost all $t \in I_{t_f}$ and terminal condition: $\eta^*(t_f) = 0$ with*

$$\mathcal{H}(x^*(t), \dot{x}^*(t), \eta^*(t), u^*(t)) = \mathcal{M}(x^*(t), \dot{x}^*(t), \eta^*(t)) \stackrel{def}{=} \max_{u \in U}\mathcal{H}(x^*(t), \dot{x}^*(t), \eta^*(t), u) \tag{5.52}$$

*for almost all $t \in I_{t_f}$.*

Evidently, (5.1), (5.48) and (5.51) with terminal condition $\eta(t_f) = 0$ and initial condition $(x, y)(0) = (x_o, 0)$ along with (5.52) constitute a nonlinear TPBVP for which the optimal time-dependent path $(x^*, y^*, \eta^*)(\cdot)$ must satisfy. For the special case where $u \in U = U_1$, Theorem 5.6 implies that $u^*(t)$ has the form given by (5.35).

The main difficulties in applying Theorems 5.4–5.6 to concrete problems lies in the fact that for most surfaces $G_f$ derived from physical situations, the visible sets $\mathcal{V}((x, f_{h_v}))$ for $x \in \Omega$ cannot be expressed analytically in terms of $x$. Consequently, the Gateaux differential in (5.47), and the partial derivatives of $\mu_2\{\Pi_\Omega \mathcal{V}((x, f_{h_v}(x)))\}$ with respect to $x_i$ in (5.32) and (5.51) cannot be readily computed. In what follows, we shall consider approximations of Problems 5.1 and 5.2, and develop numerical algorithms for their solutions.

### 5.1.4 Numerical Algorithms

To facilitate the development of numerical algorithms for the optimal motion planning problems, a mesh on $\Omega$ using standard methods such as Delaunay triangulation is established. Then, the surface under observation $G_f$ is approximated by a polyhedral surface $\hat{G}_f$ such as a surface formed by triangular patches. In physical situations, the function $f = f(x)$ is usually given in the form of numerical data. An approximate spatial profile of $G_f$ can be obtained by interpolation of the given numerical data. In what follows, numerical algorithms are developed for the approximations of Problems 5.1 and 5.2 that make use of the numerical data directly.

First, consider the Minimum-Time Total-Visibility Problem 5.1. Suppose that on $\Omega$, a mesh $M_N = \{x^{(k)} \in \Omega : k = 1, \ldots, N\}$ along with the approximate surface $\hat{G}_f$ formed by triangular patches have been constructed. Then, for each point $x^{(k)} \in M_N$, the corresponding visible set $\mathcal{V}((x^{(k)}, f_{h_v}(x^{(k)})))$ and the measure of its projection onto $\Omega$ (i.e. $\mu\{\mathcal{V}((x^{(k)}, f_{h_v}(x^{(k)})))\}$) can be computed. Let $x^{(1)} \in \Omega$ be the initial point. An *admissible path* $\hat{\Gamma}$ *initiating from* $(x^{(1)}, f(x^{(1)}))$ corresponds to a Jordan arc starting from the point $(x^{(1)}, f(x^{(1)})) \in \hat{G}_f$ formed by the edges of the triangular patches. Let $\hat{\mathcal{P}}$ denote the set of all admissible paths $\hat{\Gamma}$ initiating from $(x^{(1)}, f(x^{(1)})) \in \hat{G}_f$. An admissible path $\hat{\Gamma}_i \in \hat{\mathcal{P}}$ can be uniquely identified by a string $S_i = (\sigma_{i1}, \ldots, \sigma_{iK_i})$ with distinct entries, where $\sigma_{ij} \in M_N$, and $\sigma_{i1} = x^{(1)}$. Also, $\sigma_{ij}$ and $\sigma_{i(j+1)}$ correspond ro a pair of adjacent mesh points sharing a common edge of some triangular patch. By computing the measure of the visible set associated with each point along an admissible path $\hat{\Gamma}_i \in \hat{\mathcal{P}}$, the measure of $\bigcup_{x^{(k)} \in S_i} \Pi_\Omega \mathcal{V}((x^{(k)}, f_{h_v}(x^{(k)})))$ can be computed. Thus, the subset $\hat{\mathcal{P}}_{TV}$ of $\hat{\mathcal{P}}$ consisting of paths $\hat{\Gamma}_i$ with total visibility can be determined.

The remaining task is to determine the time duration associated with each path $\hat{\Gamma}_i \in \hat{\mathcal{P}}_{TV}$, and select those paths with the smallest time duration. This task can be accomplished by constructing the boundaries of the projection of the attainable sets of (5.1) in $\Omega$, i.e. $\partial \Pi_\Omega K_x^+(t; s_x(0)) \cap \Omega$ for system (5.1) for various times starting from $(x(0), \dot{x}(0)) = (x^{(1)}, 0)$ at $t = 0$. For each path $\hat{\Gamma}_i \in \hat{\mathcal{P}}_{TV}$, whose projection onto $\Omega$ passes through the string $S_i$ of mesh points, we can compute the time duration for transition between $x^{(1)}$ and the terminal point $x^{(K_i)}$ by interpolation using the projections of the computed attainable sets for two successive times $t$ and $t'$ such that $x^{(K_i)} \in (\Pi_\Omega K_x^+(t; s_x(0)) \cap \Pi_\Omega K_x^+(t'; s_x(0)))$. Thus, the solution to the approximate Problem 5.1 can be determined by partially ordering the paths in $\hat{\mathcal{P}}_{TV}$ according to the length of the time-duration associated with each path.

The basic steps in the foregoing algorithm can be summarized as follows:

**Step 1:** For each $x^{(k)} \in M_N$, compute $\mathcal{V}((x^{(k)}, f_{h_v}(x^{(k)})))$ and $\mu_2\{\Pi_\Omega \mathcal{V}((x^{(k)}, f_{h_v}(x^{(k)})))\}$.

**Step 2:** Construct the set $\hat{\mathcal{P}}$ consisting of all admissible paths $\hat{\Gamma}_i$ starting from $(x^{(1)}, f(x^{(1)}))$.

**Step 3:** For each path $\hat{\Gamma}_i \in \hat{\mathcal{P}}$ identified by the string $S_i$ of mesh points $x^{(k)} \in \Omega$, compute $\mu_2\{\bigcup_{x^{(k)} \in S_i} \Pi_\Omega \mathcal{V}((x^{(k)}, f_{h_v}(x^{(k)})))\}$.

**Step 4:** Determine the set $\hat{\mathcal{P}}_{TV} = \{\hat{\Gamma}_i \in \hat{\mathcal{P}} : \mu_2\{\bigcup_{x^{(k)} \in S_i} \Pi_\Omega \mathcal{V}((x^{(k)}, f_{h_v}(x^{(k)})))\} = \mu_2(\Omega)\}$.

**Step 5:** Determine the terminal time $t_f^{(i)}$ for each path $\hat{\Gamma}_i \in \hat{\mathcal{P}}_{TV}$.

**Step 6:** Find the paths $\hat{\Gamma}_i^* \in \hat{\mathcal{P}}_{TV}$ having the smallest terminal time $t_f^*$.

**Step 7:** Determine the control $u^*(\cdot)$ corresponding to each optimal path $\hat{\Gamma}_i^*$.

*Remark 5.5* Step 1 involves the computation of visible sets $\mathcal{V}((x^{(k)}, f_{h_v}(x^{(k)})))$ associated with points $x^{(k)} \in \Omega$, an NP-hard problem in computational geometry. This task can be accomplished using the algorithm developed by Balmes and Wang [4]. The complexity of that algorithm is $O(np^2)$, where $n$ and $p$ are number of observation points and the number of triangular patches respectively. This algorithm can be easily modified to take into account the limited aperture of cameras or sensors.

*Remark 5.6* In general, the optimal approximate path found in Step 6 may not be unique. Hence the corresponding optimal approximate control may not be unique either.

The foregoing algorithm can be modified to solve the approximate Problem 5.1. As before, we introduce a mesh on $\Omega$ and obtain an approximate surface $\hat{G}_f$. Next, we compute the approximate gradient of $f$ from the values of $f$ at the mesh points, and then delete all those mesh points at which the gradient constraint (5.12) is violated. Thus, we obtain a mesh for the spatial domain $\tilde{\Omega}$ as defined in Sect. 5.4. Assuming that $\hat{\Omega}$ is a simply connected set in $\mathbb{R}^2$, we can construct the set $\hat{\mathcal{P}}$ in $\tilde{\Omega}$, and follow the steps of the algorithm as before.

Now, we consider the Fixed Observation Time-interval Maximum-visibility Problem 5.2. As in Problem 5.1, we first establish a mesh $M_N = \{x^{(k)}, k = 1, \ldots, N\}$ on $\Omega$, and approximate $G_f$ by a polyhedral surface $\hat{G}_f$ formed by triangular patches. A numerical algorithm for the approximate Problem 5.2 is given by the following basic steps:

**Step 1:** Compute $\mu_2\{\Pi_\Omega \mathcal{V}((x^{(k)}, f_{h_v}(x^{(k)})) \in \hat{G}_{f_{h_v}}$ (using an algorithm such as that of Balmes and Wang [4]).

**Step 2:** Construct the set $\hat{\mathcal{P}}$ consisting of all admissible paths $\hat{\Gamma}_i$ starting from $(x^{(1)}, f(x^{(1)}))$.

**Step 3:** Determine the terminal $t_f^{(i)}$ for each path $\hat{\Gamma}_i \in \hat{\mathcal{P}}$.

**Step 4:** Construct the set $\hat{\mathcal{P}}_{FT}$ consisting of all admissible paths in $\hat{\mathcal{P}}$ having the same specified terminal time $t_f$.

**Step 5:** For each path $\hat{\Gamma}_i \in \hat{\mathcal{P}}_{FT}$ identified by the string by the string of mesh points $x^{(i)} \in \Omega$, compute $\mu_2\{\Pi_\Omega \mathcal{V}((x^{(k)}, f_{h_v}(x^{(k)})))\}$.

**Step 6:** Find an admissible path $\hat{\Gamma}_{i*} \in \hat{\mathcal{P}}_{FT}$ identified by the string of mesh points $x^{(k)} \in \Omega$, such that $\hat{J}_1(\hat{\Gamma}_{i*}) \geq \hat{J}_1(\hat{\Gamma}_i)$ for all $\Gamma_i$ for all $\hat{\Gamma}_i \in \hat{\mathcal{P}}_{FT}$, where

$$\hat{J}_1(\hat{\Gamma}_i) \stackrel{\text{def}}{=} \sum_{x^{(k)} \in \hat{\Gamma}_i} \mu_2\{\Pi_\Omega \mathcal{V}((x^{(k)}, f_{h_v}(x^{(k)})))\}. \tag{5.53}$$

In general, the strings $S_i$'s associated with the paths in $\mathcal{P}_{FT}$ may have different lengths, although their terminal times are identical. In Step 4, the set $\hat{\mathcal{P}}_{FT}$ corresponds to the projection of the approximate attainable cone truncated at the terminal time $t_f$. The truncation involves an interpolation process discussed earlier.

The foregoing algorithm can be modified to solve the approximate Problem 5.2 with visibility functional $J_2$ by replacing $\hat{J}_1(\hat{\Gamma}_i)$ in Step 6 by

$$\hat{J}_2(\hat{\Gamma}_i) \overset{\text{def}}{=} \mu_2 \{ \bigcup_{x^{(k)} \in \hat{\Gamma}_i} \Pi_\Omega \mathcal{V}((x^{(k)}, f_{h_v}(x^{(k)}))) \}. \tag{5.54}$$

### 5.1.5  Simple Example

To illustrate the application of the foregoing results and algorithms, we consider again the observed surface $G_f \subset \mathbb{R}^3$ in Example 4.1 for which the visible sets at any point in the constant-vertical height observation platform $G_{f_{h_v}}$ can be computed analytically. Let $\Omega$ be the normalized spatial domain specified by the unit disk $\{x \in \mathbb{R}^2 : \|x\| \leq 1\}$, where $x$ has been normalized with respect to the radius $r_o$ of the actual spatial domain. The surface under observation corresponds to the graph of the real-valued function $f$ given by

$$f(x) = 1 - \|x\|^2, \quad x \in \Omega, \tag{5.55}$$

where $x = (x_1, x_2)$, and $\|x\|^2 = x_1^2 + x_2^2$. It can be verified by elementary computations that for any given $h_v > 0$, the projection of the visible set from a point $(x, f_{h_v}(x)) \in G_{f_{h_v}}$ onto $\Omega$ is simply the intersection of the unit disk with the disk centered at $x$ with radius $\sqrt{h_v}$, i.e.

$$\Pi_\Omega \mathcal{V}((x, f_{h_v}(x))) = \Omega \cap \{x' = (x_1', x_2') \in \mathbb{R}^2 : \|x - x'\| \leq \sqrt{h_v}\}. \tag{5.56}$$

Moreover, for $0 < h_v \leq 1$, the measure of $\Pi_\Omega \mathcal{V}((x, f_{h_v}(x)))$ is given by

$$\mu_2\{\Pi_\Omega \mathcal{V}((x, f_{h_v}(x)))\} = \begin{cases} h_v \pi, & \text{if } 0 \leq r \leq 1 - \sqrt{h_v}; \\ \beta(r, \tilde{r}), & \text{if } 1 - \sqrt{h_v} < r \leq 1, \end{cases} \tag{5.57}$$

where $r = \|x\|$, $\tilde{r} = (1 - h_v + r^2)/2r$, and

$$\beta(r, \tilde{r}) = \cos^{-1}(\tilde{r}) + h_v \cos^{-1}\left(\frac{r - \tilde{r}}{\sqrt{h_v}}\right) - r\sqrt{1 - \tilde{r}^2}. \tag{5.58}$$

Note that for $h_v = 1$, the surface $G_{\mu_2}$ is not smooth at the origin. Moreover, $h_v = 1$ corresponds to the critical vertical height $h_{vc}$ at the origin defined in Sect. 2.1.

Consider the Minimum-Time Total Visibility Problem 5.1. Let the control region be $U_1 \overset{\text{def}}{=} \{(u_1, u_2) \in \mathbb{R}^2 : |u_i| \leq \bar{u}_i, i = 1, 2\}$. We determine the projection of the

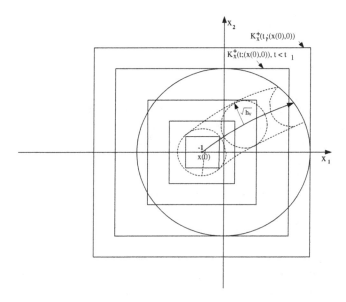

**Fig. 5.2** Projection of attainable sets and trajectories of the augmented system (5.24) onto the position space

attainable sets of (5.1) on $\Omega$ at any time $t > 0$ starting from $s_x(0) = (x(0), \dot{x}(0))$ with $x(0) = (-1, 0)$ and $\dot{x}(0) = (0, 0)$. From the solution of (5.1), we obtain

$$|x_i(t) - x_i(0)| \leq \int_0^t \frac{1}{\nu_x}(1 - \exp(-\nu_x(t - \tau)/M))|u_i(\tau)|d\tau$$

$$\leq \frac{\bar{u}_i}{\nu_x}\left(t + \frac{M}{\nu_x}(1 - \exp(-(\nu_x t/M)))\right) \tag{5.59}$$

for all $t \geq 0$ and $i = 1, 2$. Thus,

$$\Pi_\Omega K_x^+(t; (x(0), 0))$$

$$= \left\{ x \in \Omega : |x_i - x_i(0)| \leq \frac{\bar{u}_i}{\nu_x}\left(t + \frac{M}{\nu_x}(1 - \exp(-(\nu_x t/M)))\right), i = 1, 2 \right\}. \tag{5.60}$$

Evidently, there exists a finite time $t_f$ such that $\Omega \subset \Pi_\Omega K_x^+(t_f; (x(0), 0)))$. But there does not exist a time-dependent path $\Lambda = \{x_u(t; (x(0), 0)) \in \partial(\Pi_\Omega K_x^+(t; (x(0), 0)) : t \in I_{t_f}\}$ such that $G_f$ is totally visible along its corresponding path in $G_{h_v}$ i.e. $\bigcup_{x \in \Lambda} \Pi_\Omega \mathcal{V}((x, f_{h_v}(x))) = \Omega$ as illustrated in Fig. 5.2.

Next, we consider the existence of paths in $\Omega$ with total visibility. For $0 < h_v < 1$, the projection of the visible set corresponding to the closed circular path $\Gamma_{h_v} = \{x \in$

$\Omega : \|x\| = (1 - \sqrt{h_v})\}$ onto $\Omega$ is the annular region:

$$\mathcal{A}_{h_v} \overset{\text{def}}{=} \bigcup_{x \in \Gamma_{h_v}} \Pi_\Omega \mathcal{V}((x, f_{h_v}(x))) = \{x \in \Omega : (1 - 2\sqrt{h_v}) \leq \|x\| \leq 1\}. \qquad (5.61)$$

When $1 - \sqrt{h_v} \leq \|x(0)\| \leq 1$, we can construct an arc $A$ generated by an admissible control that initiates from $x(0)$, and is tangent to the circular path $\Gamma_{h_v}$ moving in the clockwise direction as shown in Fig. 5.3. The projection of the visible set of $(x(0), f_{h_v}(x(0)))$ onto $\Omega$ is given by $\Omega \cap \{x \in \mathbb{R}^2 : \|x - x(0)\| \leq \sqrt{h_v}\}$. Evidently, there exists a subset of the annular region $\mathcal{A}_{h_v}$ containing the boundary of $\Omega$ that is visible from the path $\tilde{\Gamma}_{h_v} \overset{\text{def}}{=} A \cup \mathcal{C}_{h_v}$, where

$$\mathcal{C}_{h_v} = \{x \in \Omega : \|x\| = (1 - \sqrt{h_v}), \theta_o \geq \theta \geq \theta_1\} \qquad (5.62)$$

with $\theta = \tan^{-1}(x_2/x_1)$; $\theta_o = \tan^{-1}(x_2(0)/x_1(0))$; $\theta_1 = \tan^{-1}(\hat{x}_2/\hat{x}_1)$, and $\hat{x} = (\hat{x}_1, \hat{x}_2)$ being the point where $\|\hat{x} - x(0)\| = \sqrt{h_v}$ and $\|\hat{x}\| = 1$. By continuing $\mathcal{C}_{h_v}$ along a spiral path about the origin, total visibility can be attained. We omit the details here. In the special case where $h_v = 1/4$, it is unnecessary to continue the path $\mathcal{C}_{h_v}$, since total visibility is already attained by the path $\tilde{\Gamma}_{h_v}$. When $\|x(0)\| < 1 - \sqrt{h_v}$, it is possible to construct a smooth arc connecting $x(0)$ and the circular path $\Gamma_{h_v}$ to attain total visibility.

To compute the trajectory and control that satisfy the necessary condition for optimality for Problem 5.1 given by Theorem 5.1, it is convenient to introduce a

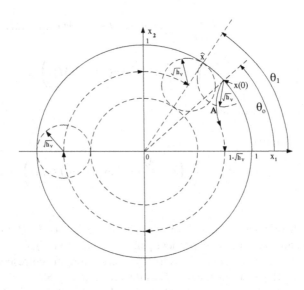

**Fig. 5.3** Construction of the arc $A$ such that a subset of the annular region $\mathcal{A}$, containing the boundary of $\Omega$ is visible from the path $\tilde{\Gamma}_{h_v} = A \cup \mathcal{C}_{h_v}$

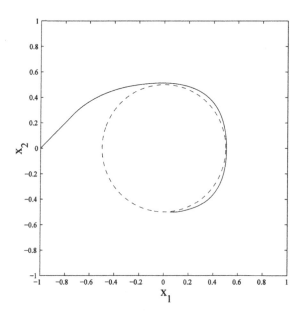

**Fig. 5.4** Projection of the computed trajectories onto the $(x_1, x_2)$-plane for Problem 5.1 with $h_v = 1/4$

normalized time $\tau = t/t_f$ and solve the TPBVP stated in Theorem 5.1 with a variable parameter $t_f$. However, even for the simple $f$ given by (5.55), the computation of $g(x)$ defined by (5.26) for any $x \in \Omega$ is a tedious task. The characterization of optimal control given by (5.35) suggests seeking "bang-bang" controls with a finite number of switchings to achieve total visibility in minimum time. Starting with the case with no switchings in both $u_1$ and $u_2$, then one switching in $u_1$ and no switching in $u_2$ etc., we obtain a trajectory in the $(x_1, x_2)$-plane with total visibility with the smallest terminal time $t_f$. Figures 5.4, 5.5 and 5.6 show the results for the case where $h_v = 1/4$. It can be seen from Fig. 5.4 that the computed trajectory in the $(x_1, x_2)$-plane approaches the circle with radius $\sqrt{h_v} = 1/2$ as fast as possible, and then stays in the neighborhood of this circle for the remaining times until total visibility is attained. At any point $x$ on this circle, $\mu_2\{\Pi_\Omega \mathcal{V}((x, f_{h_v}(x)))\}$ takes on its maximum value $\pi h_v$. Finally, we observe from Fig. 5.6 that $w = w(t)$ is indeed continuous along the computed optimal path as assumed in Theorem 5.1.

Next, consider the Fixed Observation Time-interval Maximum Visibility Problem 5.2. Here, the augmented system corresponding to (5.1) and (5.48) has the following form:

$$\frac{d}{dt}\begin{bmatrix} x \\ \dot{x} \\ y \end{bmatrix} = \begin{bmatrix} \dot{x} \\ (-\nu_x \dot{x} + u)/M \\ \mu_2\{\Pi_\Omega \mathcal{V}((x, f_{h_v}(x)))\} \end{bmatrix}, \tag{5.63}$$

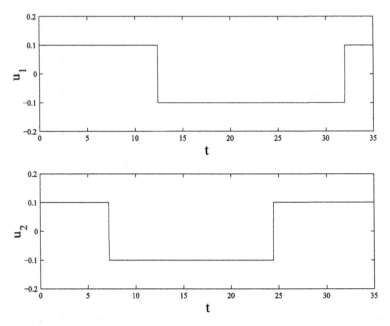

**Fig. 5.5** Computed controls $u_1 = u_1(t)$ and $u_2 = u_2(t)$ for Problem 5.1

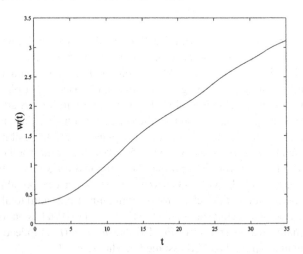

**Fig. 5.6** Time-domain plot of $w = w(t)$ along the computed optimal path for Problem 5.1

where $\mu_2\{\Pi_\Omega \mathcal{V}((x, f_{h_v}(x)))\}$ is given by (5.56). The adjoint system corresponding to (5.63) has the following explicit form:

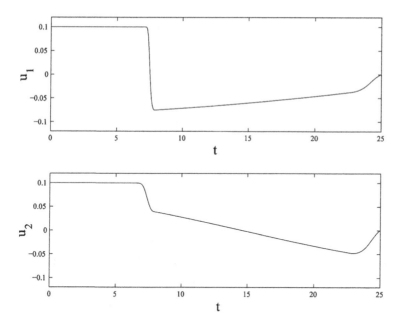

**Fig. 5.7** Computed controls $u_1 = u_1(t)$ and $u_2 = u_2(t)$ for Problem 5.2

$$\frac{d}{dt}\begin{bmatrix} \eta_1 \\ \eta_2 \\ \eta_3 \\ \eta_4 \end{bmatrix} = -\begin{bmatrix} h_1(x) \\ h_2(x) \\ \eta_1 - (\nu_x/M)\eta_3 \\ \eta_2 - (\nu_x/M)\eta_4 \end{bmatrix},$$ (5.64)

where

$$h_i(x) = \begin{cases} 0, & \text{if } 0 \le r \le 1 - \sqrt{h_v}; \\ -2x_i\sqrt{1 - \bar{r}^2}/r, & \text{if } 1 - \sqrt{h_v} < r < 1, \quad i = 1, 2. \end{cases}$$ (5.65)

Since $x \rightarrow \mu_2\{\Pi_\Omega \mathcal{V}((x, f_{h_v}(x)))\}$ given by (5.56) is $C_1$, it follows from Theorem 5.3 that the optimal control $u^*$ has the form:

$$u_1^*(t) = \bar{u}_1 \text{sgn}(\eta_3^*(t)), \quad u_2^*(t) = \bar{u}_2 \text{sgn}(\eta_4^*(t)).$$ (5.66)

Consider the nonlinear TPBVP corresponding to (5.63)–(5.66) with initial condition $(x, \dot{x}, y)(0) = (x(0), \dot{x}(0), \mu_2\{\Pi_\Omega \mathcal{V}((x(0), f_{h_v}(x(0))))\})$ and terminal condition $\eta(t_f) = 0$. Numerical solutions for this problem are obtained for specified values of the system parameters using the MATLAB algorithm "BVP4C" by Shampine et al. [5]. Figures 5.5, 5.6 and 5.7 show the results for the case where $h_v = 1/4$; $x(0) = (-1, 0)$, $\dot{x}(0) = (0.025, -0.05)$, $y(0) = \mu_2\{\Pi_\Omega \mathcal{V}((x(0), f_{h_v}(x(0))))\}$; $M = 10$ kg, $\nu_{x1} = 0.01$ N. sec./m, $\nu_{x2} = 0.02$ N.sec./m, $t_f = 25$ sec., and $\bar{u} = 0.1$ N/m. In the computation, the signum function in (5.66) was approximated

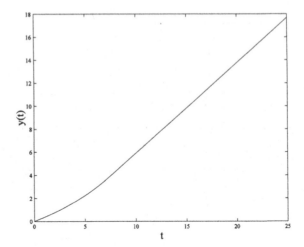

**Fig. 5.8** Computed solution to TPBVP corresponding to Problem 5.2 projected onto the $(x_1, x_2)$-plane

by $\tanh(50\eta_i^*(t)), i = 3, 4$. It can be seen from Fig. 5.8 that the main portion of the projected trajeory on $\Omega$ lies inside the disk $D$ with radius $\sqrt{h_v} = 1/2$, where $\mu_2\{\Pi_\Omega \mathcal{V}((x, f_v(x)))\}$ takes on its maximum value $\pi h_v$. This result is consistent with intuition that in order to maximize $J_1$, the trajectory should enter the disk $D$ as quickly as possible and stay inside $D$ as long as possible for the remaining times.

## 5.2 Optimal Control Problems Involving Set Measures [6]

The optimal motion planning problems for a particular system discussed in Sect. 5.1 can be regarded as special optimal control problems involving set measures for a certain class of nonlinear dynamical systems to be considered in this section. As in Sect. 5.1, let $I_{t_f}$ denote the observation or control time-interval, and $\mathcal{U}_{\mathrm{ad}}(I_{t_f})$ the set of all admissible controls $u(\cdot)$ defined on $I_{t_f}$. Let the system motion be described by

$$\dot{s} = \varphi(s) + \chi(s)u, \quad s(0) = s_o \in \Sigma, \tag{5.67}$$

where $s \rightarrow \varphi(s)$ is a given $C_1$-function on the state space $\Sigma$ into $\Sigma \subseteq \mathbb{R}^n$, whose representation with respect to a given orthonormal basis for $\Sigma$ is given by $[\varphi_1(s), \ldots, \varphi_n(s)]^T$; $u$ is the control taking its values in the control region $U$, a given nonempty compact subset of $\mathbb{R}^m$; $(s, u) \rightarrow \chi(s)u$ is a specified $C_1$-mapping on $\Sigma \times U$ into $\mathbb{R}^n$, where $\chi(s)$ may be represented with respect to suitable bases by a real $n \times m$ matrix with elements $\chi_{ij}(s)$. The set of all admissible controls defined on $I_{t_f}$ is denoted by $\mathcal{U}_{\mathrm{ad}}(I_{t_f})$. A special case of (5.67) is the linear system described by

$$\dot{s} = As + Bu, \quad s(0) = s_o \in \Sigma, \tag{5.68}$$

where $A$ and $B$ are linear transformations on the state space $\Sigma$ into $\Sigma$, and $\mathbb{R}^m$ into $\mathbb{R}^n$ represented by $n \times n$ and $n \times m$ real matrices respectively.

Let $s \to \tilde{\mathcal{V}}(s)$ be a given set-valued mapping on $\tilde{\Sigma} \to 2^{\mathbb{R}^n}$, where $\tilde{\Sigma}$ is a nonempty subset of $\Sigma$ such that $\tilde{\mathcal{V}}(s)$ is defined and compact. Moreover, $\mu_n\{\tilde{\mathcal{V}}(s)\} \le \hat{\mu}_n < \infty$ for all $s \in \tilde{\Sigma}$, where $\mu_n$ denotes the Lebesgue measure for sets in $\mathbb{R}^n$, and $\hat{\mu}_n$ is a given positive number. We assume that $\tilde{\mathcal{V}}$ is continuous with respect to the Euclidean metric $\rho_E$ and Hausdorff metric $\rho_H$. For the particular case described in (5.1), we identify $s$ with $(x_1, x_2, \dot{x}_1, \dot{x}_2)$, and $\tilde{\mathcal{V}}(s)$ with the visible set $\mathcal{V}((x_1, x_2, f_{h_v}(x_1, x_2)))$ defined for $s \in \tilde{\Sigma} \stackrel{\text{def}}{=} \{(x_1, x_2, \dot{x}_1, \dot{x}_2) \in \mathbb{R}^4 : (x_1, x_2) \in \Omega\}$.

## 5.2.1 Statement of Problems

Now, generalizations of the Optimal Motion-Planning Problems 5.1 and 5.2 can be stated as follows:

**Problem 5.1' Generalized Minimum Time Total Visibility Problem.** Let $\mathcal{U}_{\text{ad}} = \bigcup_{t_f \ge 0} \mathcal{U}_{\text{ad}}(I_{t_f})$ be the set of all admissible controls. Given $s_o \in \tilde{\Sigma}$, the initial state of (5.67) at $t = 0$, find the smallest time $t_f^* \ge 0$ and an admissible control $u^* = u^*(t)$ defined on $I_{t_f^*}$ such that its corresponding system response $s_{u^*}(t; s_o) \in \tilde{\Sigma}$ for all $t \in I_{t_f^*}$, and satisfies the terminal condition at $t_f^*$:

$$\mu_n\Big\{ \bigcup_{t \in I_{t_f^*}} \tilde{\mathcal{V}}(s_{u^*}(t; s_o)) \Big\} = c_o, \tag{5.69}$$

where $c_o$ is a specified positive constant.

**Problem 5.2' Generalized Maximum Visibility Problem with Fixed Observation Time-Interval.** Given a finite observation time interval $I_{t_f}$ and $s_o \in \tilde{\Sigma}$, the initial state of (5.67) at $t = 0$, find an admissible control $u^* = u^*(t)$ defined on $I_{t_f}$, and its corresponding response $s_{u^*}(t; s_o) \in \tilde{\Sigma}$ for all $t \in I_{t_f}$ such that the functional defined by

$$\tilde{J}_1(u) = \int_0^{t_f} \mu_n\{\tilde{\mathcal{V}}(s_u(t; s_o))\} dt \tag{5.70}$$

satisfies $\tilde{J}_1(u^*) \ge \tilde{J}_1(u)$ for all $u(\cdot) \in \mathcal{U}_{\text{ad}}(I_{t_f})$.

### 5.2.2  Optimality Conditions

As in the special case described in Sect. 5.1, optimality conditions associated with Problem 5.1' for (5.67) can be derived under the assumption that a solution exists. Let

$$w_u(t; s_o) \overset{\text{def}}{=} \mu_n \Big\{ \bigcup_{\tau \in I_t} \tilde{\mathcal{V}}(s_u(\tau; s_o)) \Big\} \tag{5.71}$$

with $w_u(0; s_o) = \mu_n\{\tilde{\mathcal{V}}(s_o)\}$. Problem 5.1' corresponds to finding the smallest time $t_f^* \in \mathbb{R}^+ = [0, \infty[$ and an admissible control $u^* = u^*(t)$ defined on $I_{t_f^*}$ such that $w_{u^*}(t_f^*; s_o) = c_o$.

Let $K_w^+(t; s_o)$ denote the attainable set corresponding to $w$ at time $t$ defined by

$$K_w^+(t; s_o) \overset{\text{def}}{=} \{ w_u(t; s_o) \in \mathbb{R}^+ : u(\cdot) \in \mathcal{U}_{\text{ad}}(I_t) \} \tag{5.72}$$

with $K_w^+(0; s_o)$ being the singleton $\mu_n\{\tilde{\mathcal{V}}(s_o)\}$. Using the same reasoning used in proving Proposition 5.2, Theorems 5.1 and 5.3, similar results for the generalized system (5.67) can be established:

**Proposition 5.2'**  *Assume that an optimal control $u^*(\cdot)$ defined on $I_{t_f^*}$ for Problem 5.1' exists. Then, $w_{u^*}(t_f^*; s_o)$ is the upper boundary point of $co(K_w^+(t_f^*; s_o))$, the convex hull of $K_w^+(t_f^*; s_o)$.*

**Theorem 5.1'**  *Suppose there exists an admissible control $\tilde{u}(\cdot)$ for (5.67) defined on some finite time interval $I_{\tilde{t}_f}$ with corresponding motion $s_{\tilde{u}} = s_{\tilde{u}}(t; s_o)$, such that*

$$\mu_n \Big\{ \bigcup_{t \in I_{\tilde{t}_f}} \tilde{\mathcal{V}}(s_{\tilde{u}}(t; s_o)) \Big\} = c_o.$$

*If the mapping $s \to \tilde{\mathcal{V}}(s)$ on $\tilde{\Sigma}$ into $2^{\mathbb{R}^n}$ is continuous with respect to the Euclidean metric $\rho_E$ and Hausdorff metric $\rho_H$, then there exists a minimum time $t_f^* \leq \tilde{t}_f$ and an optimal control $u^*(\cdot) \in \mathcal{U}_{\text{ad}}(I_{t_f^*})$ such that*

$$\mu_n \Big\{ \bigcup_{t \in I_{t_f^*}} \tilde{\mathcal{V}}((s_{u^*}(t; s_o))) \Big\} = c_o. \tag{5.73}$$

**Theorem 5.3'**  *If $w_u = w_u(t; s_o)$ is continuous in $t$ for any admissible control $u(\cdot)$, then the time-optimal control $u^*$ defined on $I_{t_f^*}$ for Problem 5.1' is an extremal control that steers $w_u(0; s_o)$ to the boundary of the attainable set corresponding to $w$ at time $t_f^*$, i.e. $w_u(t_f^*; s_o) \in \partial K_w^+(t_f^*; s_o)$.*

Next, we shall develop optimality conditions for Problem 5.1' that are analogous to Theorem 5.4.

Here, we consider the augmented system associated with (5.67):

$$\frac{d}{dt}\begin{bmatrix} s \\ w \end{bmatrix} = \begin{bmatrix} \varphi(s) + \chi(s)u \\ g(s) \end{bmatrix},$$ (5.74)

where

$$g(s_u(t; s_o)) \overset{\text{def}}{=} \lim_{\delta t \to 0^+} \sup \frac{1}{\delta t}\left[ \mu_n\{ \bigcup_{\tau \in I_{t+\delta t}} \tilde{\mathcal{V}}(s_u(\tau; s_o))\} - \mu_n\{ \bigcup_{\tau \in I_t} \tilde{\mathcal{V}}(s_u(\tau; s_o))\} \right],$$ (5.75)

with initial state at $t = 0$ given by $s_{(x,w)}(0) = (s_o, \mu_n\{\tilde{V}(s_o)\}) \in \tilde{\Sigma} \times \mathbb{R}^+$.
Let the Hamiltonian for (5.74) be defined by:

$$\mathcal{H}(s, w, \eta, u) = \tilde{\eta}^T(\varphi(s) + \chi(s)u) + \eta_{n+1}g(s),$$ (5.76)

where $\eta = (\tilde{\eta}^T, \eta_{n+1})^T$ corresponds to the state of the adjoint system:

$$\dot{\eta} = -\nabla_{(s,w)}\mathcal{H},$$ (5.77)

where $\tilde{\eta} = [\eta_1, \ldots, \eta_n]^T$.

If the real-valued function $s \to g(s)$ on $\tilde{\Sigma} \to \mathbb{R}^+$ is smooth, then the following necessary condition for optimality follows from the Pontryagin Maximum Principle [1–3]:

**Theorem 5.4'** *Suppose that the function $s \to g(s)$ on $\tilde{\Sigma} \to \mathbb{R}^+$ is $C_1$. Let $u^* = u^*(t)$ be an optimal control for Problem 5.1' with corresponding response $(s^*, w^*) = (s^*(t), w^*(t))$. Then there exists an absolutely continuous function $\eta^* = \eta^*(t)$ satisfying the adjoint equation (5.77) given explicitly by*

$$\frac{d\eta}{dt} = -\begin{bmatrix} \sum_{i=1}^n \tilde{\eta}_i \left( \nabla_s \varphi_i(s) + \sum_{j=1}^m \nabla_s \chi_{ij}(s)u_j \right) + \eta_{n+1}\nabla_s g(s) \\ 0 \end{bmatrix}$$ (5.78)

*for almost all $t \in I_{t_f}$ where*

$$\mathcal{H}(s^*(t), w^*(t), \eta^*(t), u^*(t))$$
$$= \mathcal{M}(s^*(t), w^*(t), \eta^*(t)) \overset{\text{def}}{=} \max_{u \in U} \mathcal{H}(s^*(t), w^*(t), \eta^*(t), u)$$ (5.79)

*for almost all $t \in I_{t_f}$. Moreover,*

$$\mathcal{M}(s^*(t), w^*(t), \eta^*(t)) \equiv 0 \text{ on } I_{t_f},$$ (5.80)

*and the transversality condition: $-\eta(t_f^*) = (0, \kappa)^T$ for some $\kappa > 0$ (the normal to the target set $\mathcal{T}$ at $(s^*, w^*)(t_f^*)$ in the positive $w$-direction), or*

$$\tilde{\eta}^*(t_f^*) = 0, \quad \eta_{n+1}^*(t_f^*) = -\kappa \tag{5.81}$$

*is satisfied.*

In view of Theorem 5.5' and (5.76), the optimal control $u^*$ is a function of $\eta_3^*$ and $\eta_4^*$ only. For the special case where the control region $U = U_1$, $u^*(t)$ takes on the form:

$$u_i^*(t) = \bar{u}_i \mathrm{sgn}(\eta_{i+2}^*(t)), \quad i = 1, 2. \tag{5.82}$$

Equations (5.74), (5.76)–(5.82) with terminal condition $w(t_f) = \mu_2\{\Omega\}$, initial condition $(s, w)(0) = (s_o, \mu_2\{\Pi_\Omega \mathcal{V}((s_o))\})$, and transversality condition (5.81) constitute a family of two-point-boundary-value problems (TPBVP) with the terminal time $t_f$ as a variable parameter. The optimal trajectory $(x^*, \dot{x}^*, w^*)(\cdot)$ is a solution of the TPBVP with the smallest terminal time $t_f^*$.

Now, consider Problem 5.2' with the assumption that an optimal control $u^* = u^*(t)$ defined on $I_{t_f}$ exists. Let $\delta u$ be a control perturbation such that $u = u^* + \delta u$ is admissible. Let the solutions of (5.67) at time $t$ corresponding to $u$ and $u^*$, and the same initial state $s_o$ be denoted by $s_u(t)$ and $s_{u^*}(t)$ respectively. To derive optimality conditions, consider

$$
\Delta \tilde{J}_1 \overset{\text{def}}{=} \tilde{J}_1(u^*) - \tilde{J}_1(u^* + \delta u)
$$

$$
= \int_0^{t_f} (\mu_n\{\tilde{\mathcal{V}}(s_{u^*}(t))\} - \mu_n\{\tilde{\mathcal{V}}(s_{u^*+\delta u}(t))\})dt, \tag{5.83}
$$

Thus, a sufficient but not necessary condition for optimality is given by

$$\mu_n\{\tilde{\mathcal{V}}(s_{u^*}(t))\} \geq \mu_n\{\tilde{\mathcal{V}}(s_{u^*+\delta u}(t))\} \tag{5.84}$$

for almost all $t \in I_{t_f}$, and all admissible $u^* + \delta u$.

Now, consider perturbed admissible controls of the form $u^* + \alpha\delta u$, where $\delta u$ is a given control perturbation, and $0 \leq \alpha < 1$. If the real-valued function $s \rightarrow \mu_n\{\tilde{\mathcal{V}}(s)\}$ on $\mathbb{R}^n \rightarrow \mathbb{R}^+$ is $C_1$, then the Gateaux differential of $\tilde{J}_1$ at $s_{u^*}(\cdot; s_o)$ with increment $\delta s(\cdot)$ exists. Thus, we have the following necessary condition for optimality:

**Theorem 5.5'** *Suppose that an optimal control* $u^* = u^*(t)$ *defined on* $I_{t_f}$ *for Problem 5.2' exists, and the function* $s \rightarrow \mu_n\{\tilde{\mathcal{V}}(s)\}$ *on* $\tilde{\Sigma} \rightarrow \mathbb{R}^+$ *is* $C_1$. *Then* $u^*(\cdot)$ *must satisfy the following variational inequality:*

$$
D\tilde{J}_1(u^*; \delta u) = \int_0^{t_f} \lim_{\alpha \to 0} \frac{1}{\alpha} (\mu_n\{\tilde{\mathcal{V}}(s_{u^*+\alpha\delta u}(t; s_o))^c \cap \tilde{\mathcal{V}}(s_{u^*}(t; s_o))\}
$$

$$
- \mu_n\{\tilde{\mathcal{V}}(s_{u^*}(t; s_o))^c \cap \tilde{\mathcal{V}}(s_{u^*+\alpha\delta u}(t; s_o))\})dt \geq 0 \tag{5.85}
$$

*for all admissible $u^* + \alpha\delta u$, $0 \le \alpha < 1$.*

Another necessary condition for optimality can be obtained by introducing a new state variable $y$. The evolution of $y(t)$ with time $t$ is described by

$$\dot{y}(t) = \mu_n\{\tilde{\mathcal{V}}(s_u(t))\}, \quad y(0) = 0. \tag{5.86}$$

Thus, $\tilde{J}_1(u) = y(t_f)$. Let the Hamiltonian associated with the augmented system (5.92) and (5.87) be defined by:

$$\mathcal{H}(s, \eta, u) = \mu_n\{\tilde{\mathcal{V}}(s)\} + \eta^T(\varphi(s) + \chi(s)u), \tag{5.87}$$

where $\eta = (\eta_1, \ldots, \eta_n)^T$ corresponds to the state of the adjoint system:

$$\dot{\eta} = -\nabla_s\mathcal{H}, \tag{5.88}$$

given explicitly by

$$\dot{\eta} = -\nabla_s\mu\{\tilde{\mathcal{V}}(s)\} - (D_s\varphi)^T\eta - \sum_{i=1}^{n}\eta_i\left(\sum_{j=1}^{m}\nabla_s\chi_{ij}(s)u_j\right), \tag{5.89}$$

where $D_s\varphi$ denotes the Jacobian matrix of $\varphi$ with respect to $s$. Again, if the function $s \to \mu_n\{\tilde{\mathcal{V}}(s)\}$ on $\tilde{\Sigma} \to \mathbb{R}^+$ is smooth, then a necessary condition for optimality is given by:

**Theorem 5.6'** *Suppose that the function $s \to \mu_n\{\tilde{\mathcal{V}}(s)\}$ on $\tilde{\Sigma} \to \mathbb{R}^+$ is $C_1$. Let $u^* = u^*(t)$ be an optimal control for Problem 5.2' with corresponding response $s^* = s^*(t)$. Then there exists an absolutely continuous function $\eta^* = \eta^*(t)$ satisfying (5.89) for almost all $t \in I_{t_f}$ with*

$$\mathcal{H}(s^*(t), \eta^*(t), u^*(t)) = \mathcal{M}(s^*(t), \eta^*(t)) \overset{\text{def}}{=} \max_{u \in U}\mathcal{H}(s^*(t), \eta^*(t), u). \tag{5.90}$$

Equations (5.67), (5.90) and (5.91) with initial and terminal conditions:

$$(s, y)(0) = (s_o, 0), \quad \eta(t_f) = 0 \tag{5.91}$$

constitute a nonlinear two-point-boundary-value problem for which the optimal motion $(s^*, y^*)(\cdot)$ must satisfy.

## 5.3 Asteroid Observer Optimal Motion Planning Problem

We consider again the optimal motion planning problem for an asteroid observer spacecraft described in Sect. 1.7. For illustrative purpose, we assume both the asteroid

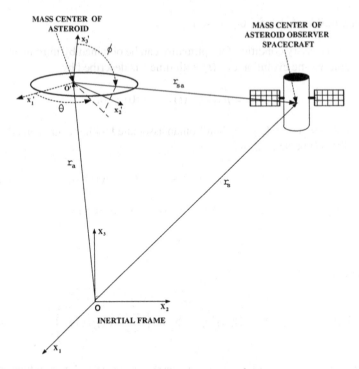

**Fig. 5.9** Coordinate system for the asteroid observer spacecraft

and observer platform surfaces are described by spheroids, and both the asteroid and observer spacecraft drift with constant velocity $\mathbf{v}_o$. Let $\mathbf{r}_{sa}(t) = \mathbf{r}_s(t) - \mathbf{r}_a(t)$, $\mathbf{r}_s(t) = \mathbf{r}(t) + \mathbf{v}_o t$, and $\mathbf{r}_a(t) = \mathbf{r}_{ao} + \mathbf{v}_o t$. Then the asteroid and the observer platform surfaces in the moving frame are given by

$$\partial \mathcal{S} = \{(r, \theta, \phi) : r(a^{-2} \sin^2 \phi + c^{-2} \cos^2 \phi)^{1/2} = 1, r \geq 0, 0 \leq \theta < 2\pi, 0 \leq \phi \leq \pi\}, \tag{5.92}$$

$$\mathcal{P}_L = \{(r, \theta, \phi) : r(a^{-2} \sin^2 \phi + c^{-2} \cos^2 \phi)^{1/2} = 1 + h/c, r \geq 0, 0 \leq \theta < 2\pi, 0 \leq \phi \leq \pi\}, \tag{5.93}$$

where $a$, $c$ and $h$ are specified positive numbers. Note that $\mathcal{P}_L$ is not a constant height observation platform.

To derive a mathematical model for the observer spacecraft motion, we consider a single spacecraft moving in the vicinity of an asteroid whose center-of-mass trajectory $\mathbf{r}_a = \mathbf{r}_a(t), t \geq 0$ relative to an inertial frame $\mathcal{F}_o$ in $\mathbb{R}^3$ is known. To model the relative motion between the spacecraft and the asteroid, it is convenient to introduce a spherical coordinate system $\mathcal{F}_a$ with its origin $O_a$ at the asteroid's center of mass (see Fig. 5.9), and with orthonormal basis $\mathcal{B} = \{e_r, e_\theta, e_\phi\}$. A point in $\mathcal{F}_a$ with respect to basis $\mathcal{B}$ is specified by the radius $r$, azimuthal angle $\phi$, and polar angle $\theta$. For a nearly ellipsoidal asteroid, its gravitational potential $\psi$ can be expressed in terms of spherical harmonics [7]:

$$\psi(r, \theta, \phi) = \frac{GM_a}{r} \left( 1 + \sum_{l=2}^{\infty} \sum_{m=0}^{l} \left( \frac{R_a}{r} \right)^l P_l^m (\cos(l\theta))(C_{lm} \sin(m\phi) + D_{lm} \cos(m\phi)) \right),$$

(5.94)

where $G$ is the universal gravitational constant; $M_a$ is the mass of the asteroid; $R_a$ is a characteristic dimension of the asteroid; $P_l^m$ is the associated Legendre polynomial; $C_{lm}$ and $D_{lm}$ are constant coefficients determined by the asteroid's mass density. For simplicity, we assume that the asteroid has no rigid-body rotational motion and the drift component of the observer spacecraft coincides with the known asteroid motion. Then only the force due to the asteroid's gravitational field is significant in modelling the asteroid-spacecraft relative motion. Thus, the equation for the spacecraft motion relative to $\mathcal{F}_a$ is given by:

$$M_s \frac{d^2 \mathbf{r}_{sa}}{dt^2} = \nabla \psi + u_s,$$

(5.95)

where $\mathbf{r}_{sa}(t) \stackrel{\text{def}}{=} \mathbf{r}_s(t) - \mathbf{r}_a(t)$; $M_s$ and $u_s$ are the mass and control force of the observer spacecraft respectively. The foregoing model will be used for deriving the necessary optimality conditions for the observer optimal motion planning problem. The equations of motion for the observer spacecraft relative to the asteroid's moving coordinate system given by (5.95) in spherical coordinates have the form:

$$M_s \ddot{\theta} = (\nabla \psi)_\theta - \frac{2M_s}{r} \dot{\theta} \dot{r} - 2M_s \dot{\theta} \dot{\phi} \cot(\phi) + \frac{u_\theta}{r \sin(\phi)},$$

(5.96)

$$M_s \ddot{\phi} = (\nabla \psi)_\phi - \frac{2M_s}{r} \dot{r} \dot{\phi} + M_s \dot{\theta}^2 \sin(\phi) \cos(\phi) + \frac{u_\phi}{r},$$

(5.97)

$$M_s \ddot{r} = (\nabla \psi)_r + M_s (r \dot{\phi}^2 + r \dot{\theta}^2 \sin^2(\phi)) + u_r,$$

(5.98)

where $r(t)$ is restricted to the observation platform or satisfies the holonomic constraint:

$$r(t) = \frac{a(c + h)}{(c^2 \sin^2(\phi(t)) + a^2 \cos^2(\phi(t)))^{1/2}}.$$

(5.99)

Thus,

$$\dot{r}(t) = \frac{a(c + h)(a^2 - c^2) \dot{\phi}(t) \sin(2\phi(t))}{2(c^2 + (a^2 - c^2) \cos^2(\phi(t)))^{3/2}}, \quad \frac{\dot{r}(t)}{r(t)} = \frac{(a^2 - c^2) \dot{\phi}(t) \sin(2\phi(t))}{2(c^2 + (a^2 - c^2) \cos^2(\phi(t)))}.$$

(5.100)

From (5.98), we have

$$u_r(t) = -(\nabla \psi)_r - M_s (r(t) \dot{\phi}^2(t) + r(t) \dot{\theta}^2(t) \sin^2(\phi(t)) - \ddot{r}(t)),$$

(5.101)

which gives the required radial control force to keep the observer spacecraft on the observation platform. Thus, the equations for the observer spacecraft relative to the asteroid's moving coordinate system reduce to (5.72) and (5.97) with $r(t)$ and $\dot{r}(t)$

given by (5.99) and (5.100) respectively. Setting $s = [s_1, \dots, s_4]^T = [\theta, \phi, \dot{\theta}, \dot{\phi}]^T$ and $u = [u_1, u_2]^T = [u_\theta, u_\phi]^T$, the observer spacecraft equations can be rewritten in the form:

$$ds/dt = \varphi(s) + \chi(s)u, \qquad (5.102)$$

where $\varphi$ and $\chi$ are given explicitly by:

$$\varphi(s) = \Big[s_3, s_4, (1/M_s)(\nabla\psi)_{s_1} - 2s_3 s_4((a^2 - c^2)\sin(s_2)\cos(s_2)/\beta(s_2) + \cot(s_2)), $$
$$(1/M_s)(\nabla\psi)_{s_2} - 2s_4^2(a^2 - c^2)\sin(s_2)\cos(s_2)/\beta(s_2) + s_3^2\sin(s_2)\cos(s_2)\Big]^T, $$
$$(5.103)$$

$$\chi(s) = \begin{bmatrix} 0 & 0 \\ 0 & 0 \\ \beta(s_2)^{1/2}/(a(c+h)M_s\sin(s_2)) & 0 \\ 0 & \beta(s_2)^{1/2}/(a(c+h)M_s) \end{bmatrix}, \qquad (5.104)$$

where $\beta(s_2) = c^2 + (a^2 - c^2)\cos^2(s_2)$. For $U = U_1$, it follows from (5.82) that the optimal controls $u_1^*(t)$ and $u_2^*(t)$ have the form:

$$u_1^*(t) = \bar{u}_1\mathrm{sgn}(\eta_3(t)/\sin(s_2(t))), \quad u_2^*(t) = \bar{u}_2\mathrm{sgn}(\eta_4(t)), \qquad (5.105)$$

provided that $\eta_3(t)/\sin(s_2(t))$ and $\eta_4(t) \neq 0$, where $\eta(t)$ is a solution of the adjoint equation:

$$\dot{\eta} = -\nabla_s(\eta^T\varphi(s)) - \nabla_s(\eta^T\chi(s)u) - \nabla_s\mu\{\mathcal{V}(\sigma(s))\}$$
$$= -\begin{bmatrix} 0 & 0 & \frac{1}{M_s}\frac{\partial}{\partial s_1}(\nabla\psi)_{s_1} & \frac{1}{M_s}\frac{\partial}{\partial s_1}(\nabla\psi)_{s_2} \\ 0 & 0 & \delta_1 & \delta_2 \\ 1 & 0 & -s_4(\kappa(s_2) + 2\cot(s_2)) & s_3\sin(2s_2) \\ 0 & 1 & -s_3(\kappa(s_2) + 2\cot(s_2)) & -2s_4\kappa(s_2) \end{bmatrix}\begin{bmatrix} \eta_1 \\ \eta_2 \\ \eta_3 \\ \eta_4 \end{bmatrix} - \begin{bmatrix} \frac{\partial\mu\{\mathcal{V}(\sigma(s))\}}{\partial s_1} \\ \frac{\partial\mu\{\mathcal{V}(\sigma(s))\}}{\partial s_2} \\ 0 \\ 0 \end{bmatrix},$$
$$(5.106)$$

where

$$\kappa(s_2) = (a^2 - c^2)\sin(2s_2)/\beta(s_2),$$

$$\delta_1 = \frac{1}{M_s}\frac{\partial}{\partial s_2}(\nabla\psi)_{s_1} - s_3 s_4\frac{d}{ds_2}(\kappa(s_2) + 2\cot(s_2)) - \frac{a\cos(s_2)u_1^*}{(c+h)M_s\beta(s_2)^{1/2}\sin^2(s_2)},$$

$$\delta_2 = \frac{1}{M_s}\frac{\partial}{\partial s_2}(\nabla \psi)_{s_2} - s_4^2 \frac{d\kappa(s_2)}{ds_2} + \cos(2s_2)s_3^2 + \frac{(c^2 - a^2)\sin(2s_2)u_2^*}{2a(c+h)M_s\beta(s_2)^{1/2}}. \quad (5.107)$$

Thus, Equation (5.102) with $u^*$ given by (5.105) and adjoint equation (5.106) along with boundary conditions:

$$s(0) = s_o, \quad (ds/dt)(0) = 0, \quad s(t_f) = w_f, \quad (ds/dt)(t_f) = 0 \quad (5.108)$$

constitute a nonlinear TPBVP for which the optimal observer spacecraft trajectory must satisfy. The visible set of any observation point $z \in \mathcal{P}_L$ can be derived from the visible set of the observation point $z = (0, a(1 + h/c), 0)$ or $(\theta, \phi) = (\pi/2, \pi/2)$ by an appropriate rotation:

$$\mathcal{V}((\pi/2, \pi/2)) = \{(r, \theta, \phi) : r \geq 0; 0 \leq \theta < 2\pi; 0 \leq \phi \leq \pi; r^2(a^{-2}\sin^2(\phi) + c^{-2}\cos^2(\phi)) = 1;$$

$$r^2(a^{-2}(\cos(\theta)\sin(\phi))^2 + c^{-2}\cos^2(\phi)) \leq (2hc + h^2)/(c+h)^2; c/(c+h) \leq ar\sin(\theta)\sin(\phi) < a\}.$$
$$(5.109)$$

The surface area of $\mathcal{V}((\pi/2, \pi/2))$ is given by

$$\mu_2\{\mathcal{V}((\pi/2, \pi/2))\} = \int_{-\Delta}^{\Delta} a\left(\pi - 2\sin^{-1}\left(\frac{c^2}{(c+h)(c^2 - x_3^2)^{1/2}}\right)\right)$$
$$\left\{1 + \frac{(a^2 - c^2)}{c^4}x_3^2\right\}^{1/2} dx_3, \quad (5.110)$$

where $\Delta = c\sqrt{(2ch + h^2)/(c+h)}$. The maximum and minimum values of $\mu\{\mathcal{V}(\cdot)\}$ occur at $(x_1', x_2', x_3') = (0, \pm(c+h), 0)$ and $(0, \pm a(1 + h/c), 0)$ respectively. Numerical solutions to the foregoing TPBVP were obtained using the MATLAB algorithm "bvp4c" by Shampine et al. [5]. For the particular case where $s_o = (3\pi/2, \pi/2, 0, 0)^T, s_f = (\pi/2, \pi/2, 0, 0)^T$ with $U = U_1, \bar{u}_1 = \bar{u}_2 = 1$ New-ton, and sufficiently large $t_f$, the optimal control steers the observer to the maximum point of $\mu\{\mathcal{V}(\cdot)\}$ as quickly as possible using "bang-bang"controls, and remains there as long as possible. This observation is consistent with the result deduced by intuitive reasoning.

Although the foregoing development reveals the nature of the solutions to the asteroid-observer optimal motion planning problem, one must take into considera-tion additional tasks associated with the problem in real-world situations. First, the asteroid motion relative to the observer spacecraft must be estimated from the mea-sured relative distance and velocity data. Moreover, the asteroid's center-of-mass must also be estimated from the measurement data including those associated with rigid-body rotation. Evidently, at any instant of time, only a portion of asteroid's surface is visible from the observer spacecraft. Therefore if the asteroid has signifi-cant rigid-body rotational motion, it may be possible to acquire sufficient measurement data for asteroid motion and total surface estimation over a finite time duration without introducing special observer spacecraft maneuvers. Thus, the observer space-craft motion optimization can be performed using the estimated asteroid surface. On the other hand, if the asteroid does not have significant rigid-body rotational motion, one may perform center-of-mass estimation and motion optimization based on partial asteroid surface measurement data, and utilize the results to determine the observer-spacecraft's optimal initial maneuvers. As the observer spacecraft acquires new asteroid surface data during the initial maneuvers, the optimization process may be repeated using the up-dated measurement data. This process may be continued until the desired objective is achieved.

## 5.4 Real-Time Observer Motion Planning

In the foregoing formulations of observer planning problems, complete knowledge of the object under observation is assumed. In real-world situations, only the union of the visible sets up to the present time $t$ is available. Therefore it is necessary to make use of the past and present observation data to plan future observer motions. Evidently, one can conceive many ways to accomplish this task. For point-observers, the computational complexity associated with surface extrapolation and motion con-trol depends on whether the observer has finite or unlimited viewing-aperture. In the latter case, the boundary of the visible set corresponding to any point-observer is determined by the tangency of rays emanated from the observer point with the surface under observation. In the former case, the boundary of the visible set is determined by the intersection of the visibility cone with the surface under observation (See Fig. 5.10). In the sequel, we shall present a few approaches for real-time observer motion-planning based on minimal use of past and present observation data for the special case described in Sect. 5.1.

To introduce the basic ideas, consider the observation of a 2D-surface $G_f$ in $\mathbb{R}^3$ using a point-observer with finite viewing-aperture moving on a constant vertical-height platform $\mathcal{P} = G_{f+h_v}$ as illustrated by Fig. 5.10. Assume that at the present time $t$, the mobile point-observer is at $z_u(t; z_o, 0) = (x_u(t; x_o, 0), f(x_u(t; x_o, 0)) + h_v) \in \mathcal{P}$ whose visible set $\mathcal{V}(z_u(t; z_o, 0))$ is indicated by the hatched surface in $G_f$. Also shown is the total visible set starting from $z_o = (x_o, f(x_o) + h_v)$ at $t = 0$ up to the present time $t$ defined by $\mathcal{V}_T(t; z_o, u, 0) \overset{\text{def}}{=} \bigcup_{0 \leq t' \leq t} \mathcal{V}(z_u(t'; z_o, 0))$. The objective is to determine the best direction for the observer movement in some sense with minimal use of the total visible set data. The approach depends on the viewing-aperture of the point-observer.

In the case where a point-observer has finite viewing aperture, its visible set at $z_u(t; z_o, 0)$ may be extrapolated to a future time $t + \Delta t$ by making use of the surface data at the boundary of the visible set $\mathcal{V}(z_u(t; z_o, 0))$. Specifically, we first compute the outward normals $\eta(x)$ along the boundary $\partial\mathcal{V}(z_u(t; z_o, 0))$ and find $\eta^*$ such that $\|\eta^*\| = \max\{\|\eta(x)\| : x \in \partial\mathcal{V}(z_u(t; z_o, 0))\}$. Then, the point-observer's $x$-position is incremented along the direction $\kappa = \eta^*/\|\eta^*\|$ to obtain the extrapolated

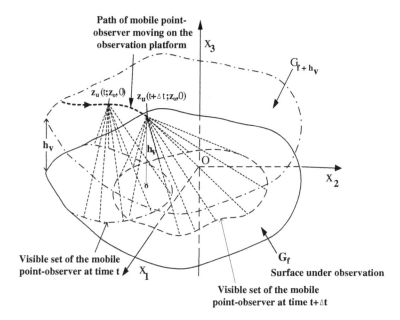

**Fig. 5.10** Observation of a surface $G_f$ by a mobile point-observer with finite viewing-aperture moving on a constant vertical-height platform $\mathcal{P} = G_{f+h_v}$

visible set $\mathcal{V}(z_u(t + \Delta t; z_o, 0))$ for sufficiently small time-increment $\Delta t$, where $z_u(t + \Delta t; z_o, 0) = (x_u(t; x_o, 0) + \kappa \Delta t, f(x_u(t; x_o, 0)) + h_v)$. The time-increment $\Delta t$ may be fixed or depends on the speed of the point-observer at time $t$. After the point-observer has moved to the new extrapolated position, the past visible set data may be revised using the visible set data at the new position.

For the case where a point-observer has unlimited viewing aperture, we determine its incremental motion by seeking a direction specified by the unit vector $\eta = (\eta_x, 0) \in \mathbb{R}^3$ that maximizes the Gateaux differential or directional derivative of $\mu_2\{\mathcal{V}_T(z_u(t; z_o, 0))\}$ defined by

$$D(\mu_2\{\mathcal{V}_T(z_u(t; z_o, 0))\}; \eta) = \lim_{\alpha \to 0^+} \frac{1}{\alpha}(\mu_2\{\mathcal{V}_T(z_u(t; z_o, 0) + \alpha\eta)\}$$
$$- \mu_2\{\mathcal{V}_T(z_u(t; z_o, 0))\}). \qquad (5.111)$$

The optimized incremental motion is given by

$$z_u(t + \Delta t; z_o, 0) = (x_u(t; x_o, 0) + \eta^*\Delta t, f(x_u(t; x_o, 0) + \eta^*\Delta t) + h_v)), \quad (5.112)$$

where $\eta^* = (\eta_x^*, 0)$ denotes the unit directional vector in $\mathbb{R}^3$ that maximizes

$$D(\mu_2\{\mathcal{V}_T(z_u(t; z_o, 0))\}; \eta).$$

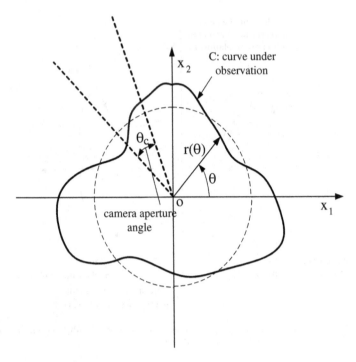

**Fig. 5.11**   Camera for observing a closed curve in the plane

## Exercises

**Ex.5.1.** Consider an observer for a spherical planet with radius $r_p$ centered at the origin of a rectangular coordinate system. The observer is represented by a point moving on a spherical platform $S$ with radius $r_o > r_p$ concentric to the planet.

(a) Assuming the absence of frictional forces, write down the equations of motion in spherical coordinates for the observer, including a control force $u$ and the necessary force for constraining the observer motion to the sphere $S$ at all times.

(b) Let $z_o \in S$ be the starting point for the observer at $t = 0$. Find the shortest path on $S$ and the corresponding observation time $t_f$ for total visibility of the planet surface.

**Ex.5.2.** A camera with its center of mass located at the origin of $\mathbb{R}^2$-plane is used to observe a closed curve $C = \{(r, \theta) : 0 \le \theta \le 2\pi, r = f(\theta)\}$, where $f$ is a specified positive $C_1$-function of $\theta$ as sketched in Fig. 5.11. The camera is capable of rotating about the axis orthogonal to the plane, and has a finite viewing aperture angle $\theta_c < \pi/4$. The camera is a rigid body whose motion is described by

$$I_o \ddot{\theta} = u, \quad \theta(0) = \theta_o, \quad \dot{\theta}(0) = \dot{\theta}_o, \tag{5.113}$$

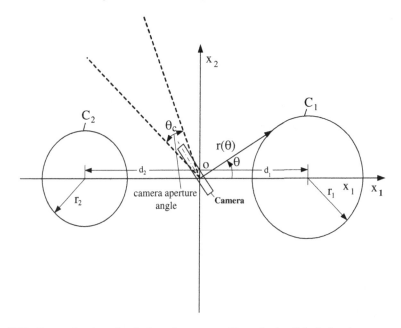

**Fig. 5.12** Camera for observing the boundary curves of two circular disks in the plane

where $I_o$ is the camera's moment of inertia about its axis of rotation. The admissible control torques $u = u(t)$ are piecewise continuous in $t$ taking their values in $U = \{u \in \mathbb{R} : |u| \le \bar{u}\}$. The problem is, given a finite observation time interval $I_{t_f}$, find an admissible control torque $u^* = u^*(t)$ defined on $I_{t_f}$ such that the total length of the observed curve is maximized.

(a) Let $\mathcal{C} = \{(r, \theta) : 0 \le \theta \le 2\pi, r = 2 + \sin(2\theta)\}$. Give an intuitive description of the optimal control torque.
(b) Derive a necessary condition for which an optimal control torque must satisfy, and give a characterization of the optimal control torque.
(c) Let the observed object be the boundaries of two closed circular disks with radii $r_1$ and $r_2$ as shown in Fig. 5.12. Find an admissible control torque $u^*(t)$ defined on $I_{t_f}$ such that the sum of the lengths of the observed curves is maximized. Repeat part (b) for this problem.

**Ex.5.3.** Consider two point-observers A and B with identical mass $m$ attached to the ends of a rigid arm with length $L$ and zero mass that can be rotated about its center by a motor with control torque $\tau$ as shown in Fig. 5.13. The The object under observation is a plane curve $f = f(x), x \in [-\pi, \pi]$. The arm center is at a fixed height $h_o$ above the origin.

(a) Let $f(x) = 1 + \cos(x) + \exp(x)$. Determine the critical height $h_v$ for a point-observer above the origin $x = 0$.

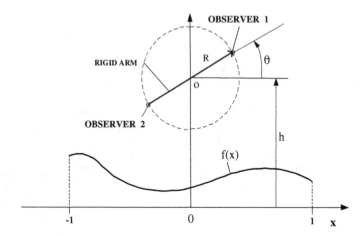

**Fig. 5.13**  Observation of a plane curve by two point-observers

(b) Find an angle $\theta_A$ such that the measure of the visible set of observer A is maximum. Repeat for observer B and observers A and B.

(c) Determine whether there exist an angle $\theta$ such that the measures of the visible sets of observer A and B are equal.

(d) Now, consider the control problem of finding a control torque $\tau = \tau(t)$ which steers the observers A and B from any initial position to the angle $\theta^*$ at which the total measure of visible sets for observers A and B is maximum.

**Ex.5.4.** Consider a three-dimensional version of Ex.5.2. Here, the camera is a rigid body with its center of mass located at the center of a closed cavity $C_v = \{(r, \theta, \phi) : |\theta| \leq \pi/2, 0 \leq \phi \leq 2\pi, r = f(\theta, \phi)\}$, where $f$ is a specified positive $C_1$-function of $\theta$ and $\phi$ (see sketch Fig. 5.14). The camera motion is described by the Euler's equation for its angular velocity $\omega$:

$$\mathcal{I}\dot{\omega} + \omega \times (\mathcal{I}\omega) = u, \tag{5.114}$$

where $\mathcal{I}$ is the tensor of inertia of the camera body. The camera attitude is described by the quaternion equation:

$$\dot{q} = D(\omega, q)q, \tag{5.115}$$

where $q = [q_1, \ldots, q_4]^T$,

$$D(\omega, q) = \begin{bmatrix} Q(q) & \omega \\ -\omega^T & 0 \end{bmatrix}, \qquad Q(q) = \begin{bmatrix} 0 & -q_3 & q_2 \\ q_3 & 0 & -q_1 \\ -q_2 & q_1 & 0 \end{bmatrix}. \tag{5.116}$$

Let the admissible control torques $u = (u_1, u_2, u_3)(t)$ be piecewise continuous functions of $t$ taking their values in $U = \{u \in \mathbb{R}^3 : |u_i| \leq \bar{u}_i, i = 1, 2, 3\}$. Assuming

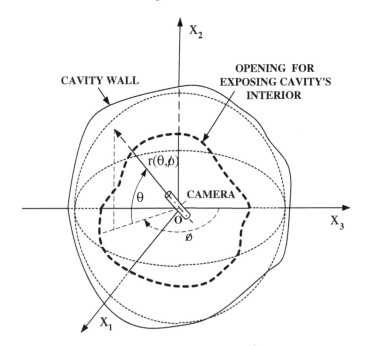

**Fig. 5.14** Camera for observing the wall of a closed cavity in $\mathbb{R}^3$

that the camera has finite viewing aperture solid angle $\alpha_v$, find an admissible control torque $u^* = u^*(t)$ such that the total area of the observed surface is maximized.

(a) Give an intuitive guess for the nature of the optimal control torque.
(b) Derive a necessary condition for which an optimal control torque must satisfy, and give a characterization of the optimal control torques.

# References

1. E.B. Lee, L. Markus, *Foundations of Optimal Control Theory* (Wiley, New York, 1967)
2. J. Macki, A. Strauss, *Introduction to Optimal Control Theory* (Springer, New York, 1982)
3. L.S. Pontryagin et al., *The Mathematical Theory of Optimal Processes* (Wiley, New York, 1962)
4. C.S. Balmes, P.K.C. Wang, Numerical Algorithms for Optimal Visibility Problems, UCLA Engineering Report ENG, pp. 00–214, June 2000
5. L.F. Shampine, J. Kierzenka, M.W. Reichelt, Solving Boundary Value Problems for Ordinary Differential Equations in MATLAB with bvp4c, Mathworks Inc, Documentation
6. P.K.C. Wang, Optimal motion planning for mobile observers based on maximum visibility. Dyn. Continuous, Discrete Impulsive Syst. Ser. B: Appl. Algorithms **11**, 313–338 (2004)
7. S.B. Broschart, D.J. Scheeres, Spacecraft Descent and Translation in the Small-body Fixed Frame,in Proceedings. AIAA/AAS Astrodynamics Specialists Meeting, Providence, R.I, Paper No. 2004–4865, p. 14, 16–19 Aug 2004

# Chapter 6
# Multiple Observer Cooperative and Non-cooperative Optimal Motion Planning

In Chap. 5, we considered various visibility-based optimal motion planning problems involving a single point-observer. In the real world, the observer may correspond to a mobile robot, spacecraft or a surveillance vehicle equipped with on-board cameras and/or sensors. Moreover, it has a solid body with finite volume. To enhance the speed of visual data acquisition, multiple observers may be used to accomplish the task. Also, multiple observers with large body-size may be used to protect an object from being observed by intruders. In what follows, we shall first consider optimal motion planning problems for multiple point-observers. These observers may cooperate or compete with each other to acquire visual information on a given object. Then, similar problems for a set of multiple observers with solid bodies will be considered. A special practically important case involves clusters of observers with solid bodies moving in prescribed formations. We shall consider various optimal motion-planning problems for this class of observers in detail, since it reveals the intrinsic complexity of these problems. To provide some insight into their solutions, a brief discussion of the solution to each problem based on intuitive reasoning will be presented first before going into mathematical details. To overcome the complexity of the general optimal motion-planning problems for multiple observers, a practical approach for obtaining sub-optimal solutions by decomposing the general problem into two groups of solvable simpler problems will be presented. Its application will be illustrated by examples from space observation systems.

## 6.1 Preliminaries

Let $I_{t_f} = [0, t_f]$ be a fixed finite observation time-interval, where $t_f < \infty$. Assume there are point-observers whose motions are described by $K$ uncoupled systems of differential equations of the form (5.67) in Chap. 5:

$$\dot{s}_{(i)} = \varphi_i(s_{(i)}) + \chi_i(s_{(i)})u_{(i)}, \quad i = 1, \ldots, K, \tag{6.1}$$

© Springer International Publishing Switzerland 2015
P.K.-C. Wang, *Visibility-based Optimal Path and Motion Planning*,
Studies in Computational Intelligence 568, DOI 10.1007/978-3-319-09779-4_6

where $s_{(i)} = [s_{(i)1}, \ldots, s_{(i)N_i}]^T$ and $u_{(i)} = [u_{(i)1}, \ldots, u_{(i)M_i}]^T$ denote the representations of the state and control for the $i$th point-observer with respect to specified bases for $\mathbb{R}^{N_i}$ and $\mathbb{R}^{M_i}$ respectively. The admissible controls $u_{(i)}(\cdot)$ are piecewise continuous functions of $t$ defined on $I_{t_f}$ taking their values in a given compact subset $U_{(i)}$ of $\mathbb{R}^{M_i}$. The set of all admissible controls $u_{(i)}(\cdot)$ is denoted by $\mathcal{U}_{\text{ad}}^{(i)}(I_{t_f})$. The *combined system* corresponding to (6.1) is described by

$$\dot{s} = \varphi(s) + \chi(s)u, \tag{6.2}$$

where $s = [s_{(1)}^T, \ldots, s_{(K)}^T]^T$, $u = [u_{(1)}^T, \ldots, u_{(K)}^T]^T$, $\varphi = [\varphi_1^T, \ldots, \varphi_K^T]^T$, and $\chi = \text{diag}[\chi_1, \ldots, \chi_K]$. As in previous chapters, let the object $\mathcal{O}$ under observation be a given compact subset of the world-space $\mathcal{W} \subset \mathbb{R}^n$, $2 \leq n \leq 3$. The projections of $s_{(i)}$ onto $\mathcal{W}$ and its corresponding visible set are denoted by $\Pi_{\mathcal{W}}s_{(i)}$ and $\mathcal{V}(\Pi_{\mathcal{W}}s_{(i)})$ respectively. Let $\mathcal{P}_i$ be the observation platform of the $i$th point-observer satisfying $\mathcal{P}_i \subset (\mathcal{O}^c \cup \partial\mathcal{O}) \subset \mathcal{W}$, where $\mathcal{O}^c$ denotes the complement of $\mathcal{O}$ relative to $\mathcal{W}$. In general, the observation platforms $\mathcal{P}_i$, $i = 1, \ldots, K$ may have non-empty intersections.

## 6.2 Cooperative Observation

One can conceive many possible schemes for cooperative observation of the object $\mathcal{O}$. We shall consider two specific problem formulations based on simple cooperative schemes. In the first scheme, we assume that before initiation of the observation task, a consensus is made among the observers to partition $\partial\mathcal{O}$ into $K$ subsets $\partial\mathcal{O}_i$, $i = 1, \ldots, K$ such that $\bigcup_{i=1}^{K} \partial\mathcal{O}_i = \partial\mathcal{O}$. The $i$th point-observer is responsible primarily for observing $\partial\mathcal{O}_i$ only. However, complete visual coverage of $\partial\mathcal{O}_i$ by the $i$th observer over the observation time interval is not required. Evidently, the set $\mathcal{V}(\Pi_{\mathcal{W}}s_{(i)}(t; u_{(i)})) \cap \partial\mathcal{O}_i$ is particularly important to the $i$th observer. Now, an optimal motion-planning problem can be posed as follows:

**Problem 6.1** Given the finite observation time-interval $I_{t_f}$ and the initial states of the point-observers at $t = 0$, $s_{(i)}(0) = s_{(i)o}$, $i = 1, \ldots, K$, find an admissible control $u = (u_{(1)}(\cdot), \ldots, u_{(K)}(\cdot))$ defined on $I_{t_f}$ such that the corresponding solution $s_{(i)}(t; u_{(i)})$ satisfies (6.1) and $\Pi_{\mathcal{W}}s_{(i)}(t; u_{(i)}) \in \mathcal{P}_i$ for all $t \in I_{t_f}$, $i = 1, \ldots, K$, and the *joint-visibility functional*

$$J_{JV}(u) \overset{\text{def}}{=} \int_0^{t_f} \sum_{i=1}^{K} \mu_n\{\mathcal{V}(\Pi_{\mathcal{W}}s_{(i)}(t; u_{(i)})) \cap \partial\mathcal{O}_i\}dt \tag{6.3}$$

takes on its maximum value.

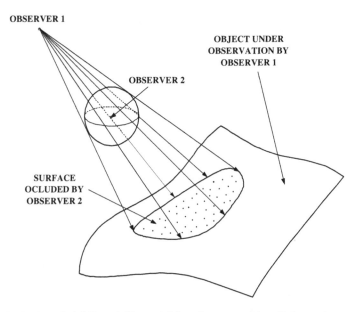

**Fig. 6.1** Reduction of visibility of *Observer 1* in reference to object $\mathcal{O}$ due to the presence of *Observer 2*

*Remark 6.1* The joint-visibility functional $J_{JV}$ involves the sum of the measures of the visible sets of $K$ point-observers. This implies that the observation data from the observers are synchronized in time over the same observation time-interval $I_{t_f}$ (an essential feature of surveillance systems for correlating time-events).

The foregoing Problem 6.1 involves point-observers. We may formulate cooperative observation problems for observers with solid bodies. Consider a pair of observers modelled by opaque closed spherical balls $\bar{\mathcal{B}}(\Pi_{\mathcal{W}}s_{(i)}(t;u_{(i)}); r_i)$, $i = 1, 2$, centered at $\Pi_{\mathcal{W}}s_{(i)}(t;u_{(i)})$ with finite radius $r_i$. Moreover, at any time $t$, it is free to move in the region $\mathcal{W} - \bigcup_{i=1,2} \bar{\mathcal{B}}(\Pi_{\mathcal{W}}s_{(i)}(t;u_{(i)}); r_i)$. The visible set of the $i$th observer at time $t$ in reference to object $\mathcal{O}$ is determined only by the intersection of the observed object (including possibly the other observer's body) with the visibility cone $\mathcal{C}$ whose vertex is at the body center $\Pi_{\mathcal{W}}s_{(i)}(t;u_{(i)})$. Figure 6.1 illustrates visibility reduction of Observer 1 with respect to object $\mathcal{O}$ due to the presence of Observer 2.

*Remark 6.2* For observers with solid bodies, their visible sets generally depend on both the position and attitude of the observer's body relative to the object $\mathcal{O}$ under observation. Thus, the complexity of the optimal motion-planning problems is greatly enhanced. Figure 6.2 illustrates this situation for observing a surface in $\mathbb{R}^3$ by an observer with a solid body. An observation path in $\mathbb{R}^3$ corresponds to that of the mass center of the observer body. Assuming that the observer has finite viewing aperture-angle, the visible set at any point along the path depends on both the distance and

**Fig. 6.2** Observation of a
surface by an observer with a
solid body

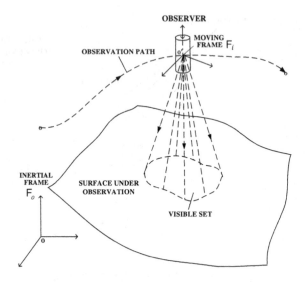

attitude of the observer's body with respect to the object's surface $\partial \mathcal{O}$. Thus the
formulation of an optimal motion-planning problem will involve seeking optimal
motions for both observer's mass center and body attitude. To facilitate the problem
formulation, we introduce the following definition of the visible set of $i$th observer
in the presence of $j$th observer centered at $z_i$ and $z_j$ respectively:

$$\mathcal{V}(z_i|z_j) \overset{\text{def}}{=} \{y \in \partial \mathcal{O} : L(z_i, y) \subset \mathcal{W} \text{ and } L(z_i, y) \cap (\mathcal{B}(z_j; r_j) \cup \text{int}(\mathcal{O})) = \phi\},$$

$$(6.4)$$

where $L(z_i, y)$ denotes the line segment joining $z_i$ and $y$; and $\mathcal{B}(z_j; r_j)$ the open
ball centered at $z_j$ with radius $r_j$. If $L(z_i, y) \cap \mathcal{B}(z_j; r_j) = \phi$ for any $y \in \partial \mathcal{O}$, then
$\mathcal{V}(z_i|z_j) = \mathcal{V}(z_i)$. Note that, in general, the visibility of an observer may be reduced
by the presence of more than one other observers. Evidently, in the formulation of
optimal motion planning problems, one must consider all cases. To avoid possible
confusion due to unavoidable complicated notations when dealing with more than
two observers, we shall limit our problem formulation to the case involving only two
observers.

Now, an optimal motion-planning problem involving a pair of observers with solid
bodies can be formulated as follows:

**Problem 6.2** Let $I_{t_f} = [0, t_f]$ be a specified finite observation time-interval, and
$s_{(i)}(0) = s_{(i)o}, i = 1, 2$ be given initial states of the body centers of the observers
at $t = 0$. Find an admissible control $u(\cdot) \in \mathcal{U}_{ad}^{(1)}(I_{t_f}) \times \mathcal{U}_{ad}^{(2)}(I_{t_f})$ such that its
corresponding motion satisfies $\Pi_{\mathcal{W}} s_{(i)}(t; u_{(i)}) \in \mathcal{P}_i, i = 1, 2$ for all $t \in I_{t_f}$, and the
visibility functional with weights:

$$J_\lambda(u) \overset{\text{def}}{=} \int\limits_0^{t_f} (\lambda \mu_n \{\mathcal{V}(\Pi_{\mathcal{W}} s_{(1)}(t; u_{(1)}) | \Pi_{\mathcal{W}} s_{(2)}(t; u_{(2)}))\}$$

$$+ (1 - \lambda)\mu_n \{\mathcal{V}(\Pi_{\mathcal{W}} s_{(2)}(t; u_{(2)}) | \Pi_{\mathcal{W}} s_{(1)}(t; u_{(1)}))\})dt \qquad (6.5)$$

takes on its maximum value, where $0 \le \lambda \le 1$.

*Remark 6.3* At the first glance, it may not be obvious that cooperation between the observers is involved in Problem 6.2. Let us examine the situation more closely. Consider first the case where $\lambda = 1$. In this case, the visible sets of Observer 2 in reference to object $\mathcal{O}$ do not contribute directly to the value of $J_\lambda$. But the visible sets of Observer 1 in reference to object $\mathcal{O}$ depend on the location of Observer 2, and affect the value of $J_\lambda$ indirectly. To maximize $J_\lambda$, one expects that Observer 2's optimal movement should reduce blocking the visibility of Observer 1. Similarly, for the case where $\lambda = 0$, the visible sets of Observer 1 only affects the value of $J_\lambda$ indirectly. To maximize $J_\lambda$, one expects that Observer 1's optimal movement should reduce blocking the visibility of Observer 2 in reference to $\mathcal{O}$. In the case where $0 < \lambda < 1$, both observers' optimal movements should reduce blocking each other's visibility as much as possible. Evidently, in all cases, the expected observer movements imply cooperation between the observers, and implicitly assume that the value of $J_\lambda$ can be increased by reducing visibility blocking. At this point, one may ask whether this assumption is always valid? We reserve this question as an exercise (see Exercise 6.3).

In what follows, we shall develop necessary optimality conditions for Optimal Motion Planning Problems 6.1 and 6.2. First, we note that Problem 6.1 involving maximizing the joint-visibility functional $J_{JV}$ is a fixed observation time-interval optimal control problem similar to Problem 5.2' discussed in Chap. 5. Thus, the following necessary conditions for optimality can be established directly from those proved in Chap. 5.

**Theorem 6.1** *Suppose that an optimal control* $u^*(\cdot) = (u^*_{(1)}, \ldots, u^*_{(K)})(\cdot)$ *defined on* $I_{t_f}$ *for Problem 6.1 exists, and the function* $u \to J_{JV}(u)$ *on* $\mathcal{U}^{(1)}_{ad}(I_{t_f}) \times \cdots \times \mathcal{U}^{(K)}_{ad}(I_{t_f})$ *into* $\mathbb{R}^+$ *is* $C_1$. *Then* $u^*(\cdot)$ *must satisfy the variational inequality:*

$$DJ_{JV}(u^*; \delta u) = \int\limits_0^{t_f} \lim_{\alpha \to 0^+} \frac{1}{\alpha} \sum_{i=1}^K (\mu_n \{(\mathcal{V}(\Pi_{\mathcal{W}} s_{(i)}(t; u^*_{(i)})$$

$$+ \alpha \delta u_{(i)}) \cap \partial \mathcal{O}_i)^c \cap \mathcal{V}(\Pi_{\mathcal{W}} s_{(i)}(t; u^*_{(i)}) \cap \partial \mathcal{O}_i))\}$$

$$- \mu_n \{\mathcal{V}(\Pi_{\mathcal{W}} s_{(i)}(t; u^*_{(i)})) \cap \partial \mathcal{O}_i)^c \cap \mathcal{V}(\Pi_{\mathcal{W}} s_{(i)}(t; u^*_{(i)})$$

$$+ \alpha \delta u_{(i)}) \cap \partial \mathcal{O}_i\})dt \ge 0 \qquad (6.6)$$

*for all admissible* $u^* + \alpha \delta u, 0 \le \alpha < 1$.

For Problem 6.2, assuming the existence of a solution, we have the following necessary condition for optimality:

**Theorem 6.2** *Suppose that an optimal control* $u^*(\cdot) = (u^*_{(1)}(\cdot), u^*_{(2)}(\cdot))$ *defined on* $I_{t_f}$ *for Problem 6.2 exists, and the function* $u \to J_\lambda(u)$ *on* $\mathcal{U}^{(1)}_{ad}(I_{t_f}) \times \mathcal{U}^{(2)}_{ad}(I_{t_f})$ *into* $\mathbb{R}^+$ *is* $C_1$. *Then* $u^*(\cdot)$ *must satisfy the variational inequality:*

$$
\begin{aligned}
DJ_\lambda(u^*; \delta u) = \int_0^{t_f} \lim_{\alpha \to 0^+} \frac{1}{\alpha} (\lambda(\mu_n\{\mathcal{V}(\Pi_\mathcal{W}s(t; u^*_{(1)})|\Pi_\mathcal{W}s(t; u^*_{(2)}))^c \cap \mathcal{V}(\Pi_\mathcal{W}s(t; u^*_{(1)}) \\
+ \alpha\delta u_{(1)})|\Pi_\mathcal{W}s(t; u^*_{(2)} + \alpha\delta u_{(2)}))\} \\
- \mu_n\{\mathcal{V}(\Pi_\mathcal{W}s(t; u^*_{(1)} + \alpha\delta u_{(1)})|\Pi_\mathcal{W}s(t; u^*_{(2)}))^c \\
\cap \mathcal{V}(\Pi_\mathcal{W}s(t; u^*_{(1)})|\Pi_\mathcal{W}s(t; u^*_{(2)}))\} \\
+ (1 - \lambda)(\mu_n\{\mathcal{V}(\Pi_\mathcal{W}s(t; u^*_{(2)})|\Pi_\mathcal{W}s(t; u^*_{(1)}))^c \cap \mathcal{V}(\Pi_\mathcal{W}s(t; u^*_{(2)} \\
+ \alpha\delta u_{(2)}))|\Pi_\mathcal{W}s(t; u^*_{(1)} + \alpha\delta u_{(1)}))) \\
- \mu_n\{\mathcal{V}(\Pi_\mathcal{W}s(t; u^*_{(2)} + \alpha\delta u_{(2)})|\Pi_\mathcal{W}s(t; u^*_{(1)} \\
+ \alpha\delta u_{(1)}))^c \cap \mathcal{V}(\Pi_\mathcal{W}s(t; u^*_{(2)})|\Pi_\mathcal{W}s(t; u^*_{(1)}))\})dt \geq 0
\end{aligned}
\tag{6.7}
$$

*for all admissible* $u^* + \alpha\delta u$, $0 \leq \alpha < 1$.

*Proof* Let $u^*(\cdot)$ be an optimal control that maximizes $J_\lambda$. Then the increment

$$
\begin{aligned}
\Delta J_\lambda \stackrel{\text{def}}{=} J_\lambda(u^* + \alpha\delta u) - J_\lambda(u^*) = \int_0^{t_f} \lambda(\mu_n\{\mathcal{V}(\Pi_\mathcal{W}s(t; u^*_{(1)}) \\
+ \alpha\delta u_{(1)})|\Pi_\mathcal{W}s(t; u^*_{(2)} + \alpha\delta u_{(2)}))\} \\
- \mu_n\{\mathcal{V}(\Pi_\mathcal{W}s(t; u^*_{(1)})|\Pi_\mathcal{W}s(t; u^*_{(2)}))\}) \\
+ (1 - \lambda)(\mu_n\{\mathcal{V}(\Pi_\mathcal{W}s(t; u^*_{(2)}) \\
+ \alpha\delta u_{(2)}))|\Pi_\mathcal{W}s(t; u^*_{(1)} + \alpha\delta u_{(1)}))\} \\
- \mu_n\{\mathcal{V}(\Pi_\mathcal{W}s(t; u^*_{(2)})|\Pi_\mathcal{W}s(t; u^*_{(1)}))\})dt \geq 0
\end{aligned}
\tag{6.8}
$$

for all admissible $u^* + \alpha\delta u$, $0 \leq \alpha < 1$. Using Lemma 4.1, the above inequality involving the Gateaux-differential of $J_\lambda$ at $u^*$ with increment $\delta u$ defined by

$$
DJ_\lambda(u^*; \delta u) \stackrel{\text{def}}{=} \lim_{\alpha \to 0^+} \frac{1}{\alpha}(J_\lambda(u^* + \alpha\delta u) - J_\lambda(u^*)),
\tag{6.9}
$$

can be rewritten in the form of the variational inequality (6.7). □

*Remark 6.4* It is evident from (6.5) that in the absence of visibility-blocking, the functional $J_\lambda$ reduces to a convex combination of the functionals of the measures

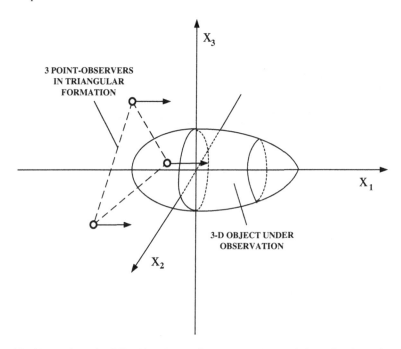

**Fig. 6.3** Observation of a 3-D object by a point-observer cluster-triad moving in a triangular formation

of the visible sets of the individual observers. For the case with visibility-blocking, the determination of the measure of relevant visible sets in the variational inequality (6.7) is generally a computational intensive task. Efficient algorithms for computing these visible sets must be developed before the variational inequality can be used effectively.

### 6.2.1 Observer-Clusters

In wide-area surveillance and mapping of terrestrial and planetary surfaces, and observation of 3-D objects such as asteroids described in Chap. 1, the observation capabilities of multiple observers can be enhanced by grouping them into clusters moving in prescribed formations and performing tasks in a cooperative manner. Recently, a number of studies have been made on the development of formation control and navigation strategies for multiple mobile robots [1] and spacecraft [2–4]. In what follows, we shall adopt some of the approaches in these works to the control of observer-clusters.

To fix ideas, consider first an observer-cluster consisting of a finite set of point-observers for observing an object in the three-dimensional Euclidean space $\mathbb{R}^3$. Figure 6.3 illustrates a 3-D object $\mathcal{O}$ under observation by a point-observer cluster

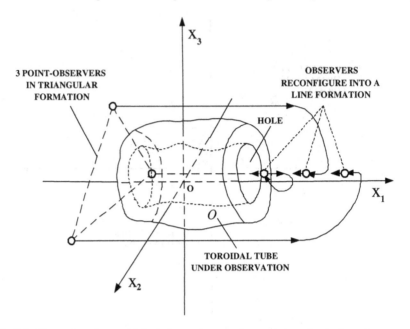

**Fig. 6.4** Observation of a *toroidal tube* by an observer-cluster triad moving in formation

triad moving in a triangular formation. The arrows indicate the directions of motion. The observers may communicate with each other during observation to minimize redundancy of observation data and/or to control their relative positions and attitudes. When the observed object $\mathcal{O}$ is not simply connected such as the toroidal tube shown in Fig. 6.4, the point-observers may reconfigure themselves from a triangular formation to a line formation so that the complete visual coverage of the surface of $\mathcal{O}$ can be attained.

Let $I_{t_f} = [0, t_f]$ be the observation time-interval, where $t_f < \infty$ is either fixed or variable. Assume there are $K$ point-observers whose motions are modelled by $K$ uncoupled systems of differential equations in the form (6.1), i.e.

$$\dot{s}_{(i)} = \varphi_i(s_{(i)}) + \chi_i(s_{(i)})u_{(i)}(t), \quad i = 1, \ldots, K, \tag{6.10}$$

where $u_{(i)} = u_{(i)}(t)$ is the control for the $i$th observer taking its values in the compact set $U_{(i)} \subset \mathbb{R}^{M_i}$. Let $s_{(i)}(t; u_{(i)})$ denote the state of the $i$th observer at time $t$ with control $u_{(i)}$, and its projection onto the world space $\mathcal{W}$ be denoted by $\Pi_{\mathcal{W}} s_{(i)}(t; u_{(i)})$. There are many schemes for forming observer-clusters. A few simple schemes are described in the sequel:

**(i) Nearest-Neighbor Tracking**

Here, an observer, say the first one, is assigned as the *leader* or *cluster reference*. Its main function is to provide a cluster reference motion whose projection onto the world space $\mathcal{W}$ is denoted by $\mathcal{M}^c = \{\Pi_{\mathcal{W}} s_{(1)}(t; u_{(1)}); t \in I_{t_f}\}$. The desired motion for the second observer (*follower*) is specified by $\{s_{(2)}^d(t); t \in I_{t_f}\}$ whose projection onto the world space $\mathcal{W}$ is given by

$$\Pi_{\mathcal{W}} s_{(2)}^d(t) = \Pi_{\mathcal{W}} s_{(1)}(t; u_{(1)}) + d_{(2)}(t), \quad t \in I_{t_f}, \qquad (6.11)$$

where $d_{(2)}(t)$ is a specified nonzero deviation vector defined for all $t \in I_{t_f}$. Moreover, to avoid the possibility of collision between the two observers, we impose the constraint $\|d_{(2)}(t)\| > \rho_1 + \rho_2$ for all $t \in I_{t_f}$, where $\rho_i$ denotes the radius of the ball containing $i$th observer. The second observer tracks the motion of the leader such that the norm of the tracking error

$$\varepsilon_{(2)}(t) \overset{\text{def}}{=} \Pi_{\mathcal{W}}(s_{(2)}^d(t) - s_{(2)}(t; u_{(2)})) \qquad (6.12)$$

is within a specified bound. The desired motion for the $i$th observer, $i > 2$, can take on many forms. Typical forms are

$$\Pi_{\mathcal{W}} s_{(i)}^d(t) = \Pi_{\mathcal{W}} s_{(1)}(t; u_{(1)}) + d_{(i)}(t), \qquad (6.13)$$

and

$$\Pi_{\mathcal{W}} s_{(i)}^d(t) = \Pi_{\mathcal{W}} s_{(i-1)}^d(t) + d_{(i)}(t) = \Pi_{\mathcal{W}} s_{(1)}(t; u_{(1)}) + \sum_{j=2}^{i} d_{(j)}(t), \qquad (6.14)$$

where $d_{(i)}(t)$ is a specified nonzero deviation vector having properties similar to those of $d_{(2)}(t)$. Here, the tracking error is defined by

$$\varepsilon_{(i)}(t) \overset{\text{def}}{=} \Pi_{\mathcal{W}}(s_{(i)}^d(t) - s_{(i)}(t; u_{(i)})). \qquad (6.15)$$

Thus, the set $C^d(t) = \{s_{(i)}^d(t), i = 1, ..., K\}$ defines a desired or reference pattern for the observer-cluster at time $t$. The actual pattern of the cluster-formation at time $t$ is specified by

$$C(t) = \{\Pi_{\mathcal{W}} s_{(1)}(t; u_{(1)}), \ldots, \Pi_{\mathcal{W}} s_{(K)}(t; u_{(K)})\}, \quad t \in I_{t_f}. \qquad (6.16)$$

Figure 6.5 shows typical cluster-formation patterns generated by (6.13) and (6.14) .

(ii) **Multi-neighbor Tracking**

Consider a cluster of $K$ observers with $K \geq 3$. Let the first and $K$th observers be designated as leaders or references for the cluster. Their motions $\mathcal{M}_1^c = \{s_{(1)}(t; u_{(1)}); t \in I_{(t_f)}\}$ and $\mathcal{M}_K^c = \{s_{(K)}(t; u_{(K)}); t \in I_{t_f}\}$ are taken as reference motions for the cluster. Here, the desired position of the $i$th observer, $i \neq 1, K$, in the world space $\mathcal{W}$ at any time $t \in I_{t_f}$ may be taken as the median of the positions of the nearest two neighbors at time $t$, i.e.

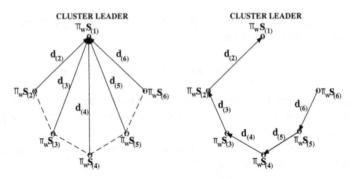

**Fig. 6.5** Typical cluster formation patterns generated by (6.13) and (6.14)

$$\Pi_{\mathcal{W}} s_{(i)}^{d}(t) = \Pi_{\mathcal{W}}(s_{(i+1)}(t; u_{(i+1)}) + s_{(i-1)}(t; u_{(i-1)}))/2. \qquad (6.17)$$

### (iii) Inertially Referenced Movements

Here, the cluster reference formation pattern $\mathcal{C}_{R_o}$ is generated by assigning a desired motion $\{s_{(i)}^{d}(t); t \in I_{t_f}\}$ relative to an inertial frame for each observer. As in Nearest-Neighbor Tracking, we require that $\|s_{(i)}^{d}(t) - s_{(i-1)}^{d}(t)\| > \rho_i + \rho_{i-1}$ for all $t \in I_{t_f}$ so as to avoid the possibility of collisions between the $i$th and $(i - 1)$th observers. Thus, each observer tries to move along its own path without any knowledge of the motions of the remaining observers in the cluster.

### (iv) Mixed Multi-neighbor and Inertially Referenced Movements

Let $\mathcal{C}_i$ be the visible cone for the $i$th observer. Then the visible cone for the cluster with respect to a given object $\mathcal{O}$ under observation is simply the union of the visible cones $\mathcal{C}_i, i = 1, \ldots, K$ with respect to $\mathcal{O}$. Assuming that the point-observers in the cluster are able to maintain the desired formation pattern at all times, one may formulate the optimal motion planning problem for the observer-cluster by regarding it as a single entity. The reference-observer determines and executes the optimal motion for the cluster, and the remaining observers tracks and follows the motion of the reference-observer. This corresponds to the "follow-the-leader" approach for cluster formation keeping [1].

The formulation of optimal motion-planning problems for an observer-cluster generally involves the development of appropriate controls for formation-keeping, attitude synchronization, formation-pattern reconfiguration and observer-cluster observation trajectory optimization based on maximum visibility of the entire observer-cluster. Evidently, taking into consideration of all the above-mentioned factors in the problem formulation leads to a highly complex optimization problem which cannot be readily solved for real-time realization of the controls. Therefore we propose a more practical approach to the optimal motion planning problem. The basic idea here is to decompose the general problem into two groups of simpler problems. The first group consists of formation-keeping, attitude synchronization and formation-pattern reconfiguration problems without considering various visibility issues. The

**Fig. 6.6** Sketch of a spacecraft-triad interferometer

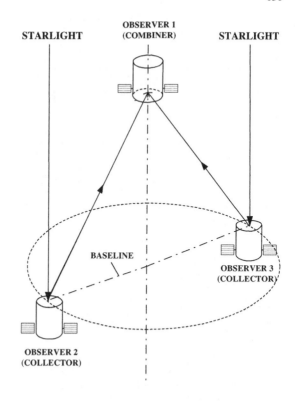

OBSERVER 1
(COMBINER)

STARLIGHT

STARLIGHT

BASELINE

OBSERVER 3
(COLLECTOR)

OBSERVER 2
(COLLECTOR)

second group consists of the optimal observer motion-planning problem assuming perfect cluster formation-keeping and attitude synchronization so that the observer-cluster can be regarded as a single entity. The proposed approach will be discussed with the aid of an example from space observation systems, namely, a spacecraft-triad long-baseline interferometer. First, a brief description of the spacecraft-triad interferometer will be presented. Then, a dynamic model for the translational and rigid-body rotational motions of the spacecraft-triad will be developed and followed by the derivation of control laws for formation-keeping, attitude synchronization and formation-pattern reconfiguration. Having obtained solutions to the problems in the first group, attention will be devoted to the problem in the second group dealing with the formulation of optimal-motion planning problem based on maximum visibility of the centralized observer-cluster.

### 6.2.1.1 Spacecraft-Triad Interferometer

A simple example of an observer-cluster is a spacecraft-triad long-baseline interferometer such as the proposed NASA DS-3 interferometer [5–7] illustrated by Fig. 6.6. Here, each observer corresponds to a spacecraft (referred to hereafter as an *observer-*

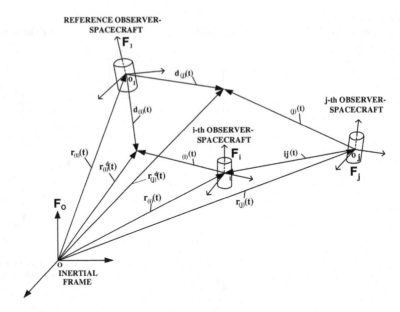

**Fig. 6.7** Sketch of an observer-spacecraft cluster and the spacecraft coordinate frames

*spacecraft*) placed at a vertex of an equilateral triangle. The first observer-spacecraft is the *combiner* and the remaining ones are the *collectors* equipped with mirrors for collecting light from a distant star. The light reflected from the the collectors' mirrors impinges on the optical sensor of the combiner. Initial interferometer operation requires keeping the observer-spacecraft in a triangular formation-pattern in world space $\mathcal{W} = \mathbb{R}^3$ with specified separation distances. Then, the attitudes of the observer-spacecraft and collector mirrors are adjusted so that the reflected light from the collectors impinges on the combiner's optical sensor with position accuracy in the sensor plane on the order of fraction of the optical wavelength. The final objective is to reconstruct the star's image that calls for gathering interferometry data from various formation orientations by slewing the observer-cluster from one orientation to another while maintaining the triangular cluster-formation pattern in a specified orbit. In the sequel, we shall use this example to clarify the development of various controls for achieving the desired objective of the observer-cluster.

The first task in our proposed approach is to derive a mathematical model describing the translational and rotational motions of the spacecraft-triad interferometer. We assume each observer-spacecraft has a rigid body with fixed center of mass. Let $\mathcal{F}_o$ denote the inertial frame with origin $O \in \mathbb{R}^3$, and $\mathcal{F}_i$ a moving body frame whose origin $o_i$ is at the mass center of the $i$th observer-spacecraft (see Fig. 6.7 for a sketch of the coordinate systems associated with a general observer-spacecraft cluster). For a given orthonormal basis $B_i$ for $\mathcal{F}_i$, the representation of a vector $v \in \mathbb{R}^3$ with respect to $B_i$ will be denoted by $[v]^i$. Let $r_{(i)}(t)$ denote the position of the mass center of the $i$th observer-spacecraft at time $t$ in $\mathbb{R}^3$ relative to $\mathcal{F}_o$. In the absence of gravitational field and disturbances, the evolution of $r_{(i)}(t)$ with time is governed by

$$\mathcal{D}_i^2 r_{(i)} = u_{(i)}/M_i, \tag{6.18}$$

where $u_{(i)}$ and $M_i$ denote the control force and mass of the $i$th observer-spacecraft respectively; $\mathcal{D}_i^2$ is the time-derivative operator defined by

$$\mathcal{D}_i^2 \stackrel{\text{def}}{=} \frac{^i d^2(\cdot)}{dt^2} + \frac{^i d\omega_{(i)}}{dt} \times (\cdot) + 2\omega_{(i)} \times \frac{^i d(\cdot)}{dt} + \omega_{(i)} \times (\omega_{(i)} \times \cdot), \tag{6.19}$$

where $^i d/dt$ and $^i(d^2/dt^2)$ denote the time derivative operators with respect to the body frame $\mathcal{F}_i$, and $\omega_{(i)}$ the angular velocity of $\mathcal{F}_i$ with respect to the inertial frame $\mathcal{F}_o$.

In addition to the foregoing equations for the translational motion of the $i$th observer-spacecraft, we also have the following quaternion and Euler equations describing respectively the attitude and angular velocity of the $i$th observer-spacecraft with respect to the inertial frame $\mathcal{F}_o$:

$$\frac{^o d\hat{q}_{(i)}}{dt} = (q_{(i)4}\omega_{(i)} - \omega_{(i)} \times \hat{q}_{(i)})/2, \qquad \frac{^o dq_{(i)4}}{dt} = -\langle \omega_{(i)}, \hat{q}_{(i)} \rangle/2, \tag{6.20}$$

$$\frac{^o d(\mathbf{I}_i\omega_{(i)})}{dt} \stackrel{\text{def}}{=} \mathbf{I}_i \frac{^i d\omega_{(i)}}{dt} + \omega_{(i)} \times (\mathbf{I}_i\omega_{(i)}) = \tau_{(i)}, \tag{6.21}$$

where $q_{(i)} = [[\hat{q}_{(i)}]^T, q_{(i)4}]^T, \hat{q}_{(i)} = [q_{(i)1}, q_{(i)2}, q_{(i)3}]^T$ and $\mathbf{I}_i$ is the tensor of inertia in the body frame $\mathcal{F}_i$.

The foregoing Eqs. (6.18)–(6.21) constitute the basic dynamic model to be used for the subsequent development of observer-cluster controls.

### 6.2.1.2 Cluster-Formation Keeping

For cluster-formation keeping, each observer-spacecraft can be regarded as a point-mass $M_i, i = 1, 2, 3$ at its center of mass. Its motion $r_{(i)}(t; u_{(i)}) = \Pi_{\mathcal{W}} s_{(i)}(t; u_{(i)})$ in the world space $\mathcal{W} = \mathbb{R}^3$ is governed by (6.18). To develop controls for cluster-formation keeping, it is natural to consider the relative motion between any pair of observer-spacecraft. Let $\rho_{ji} = r_{(j)} - r_{(i)}$ denote the position of the $j$th spacecraft relative to $\mathcal{F}_i$. Using (6.18) and (6.19), it can be verified that the evolution of $\rho_{ji}(t)$ with time is described by

$$\mathcal{D}_{(i)}^2 \rho_{ji} = u_{(j)}/M_j - u_{(i)}/M_i. \tag{6.22}$$

Adopting the nearest-neighbor scheme for cluster-formation described by (6.13) (see Fig. 6.5), the desired motion for the $i$th observer-spacecraft $i = 2, 3$ is specified by

$$r_{(i)}^d(t) = r_{(1)}(t) + d_{(i)}(t), \quad t \geq 0, \tag{6.23}$$

where $d_{(i)}(t) \in \mathbb{R}^3$ is a specified nonzero *deviation vector* defined for all $t \geq 0$. The $i$th observer-spacecraft tries to track the motion of the reference observer-spacecraft or leader such that the norm of the tracking error

$$\varepsilon_{(i)}(t) \overset{\text{def}}{=} r_{(i)}^d(t) - r_{(i)}(t) = \rho_{1(i)}(t) + d_{(i)}(t) \tag{6.24}$$

is as small as possible. The *desired cluster-formation pattern* at any time $t \geq 0$ is specified by the point set $C^d(t) = \{r_{(1)}(t), r_{(1)}(t) + d_{(2)}(t), r_{(1)}(t) + d_{(3)}(t)\}$. The actual cluster-formation pattern is described by the set $C(t) = \{r_{(1)}(t), r_{(2)}(t), r_{(3)}(t)\}$, where $r_{(i)}(t)$ denotes the position of the $i$th observer-spacecraft at time $t$ with respect to the inertial frame $\mathcal{F}_o$.

To derive a feedback control or control law for cluster formation-keeping, we consider the following equation for the tracking error $\varepsilon_{(i)}(t) \overset{\text{def}}{=} r_{(1)}(t) + d_{(i)} - r_{(i)}(t)$, $i = 2, 3$:

$$\mathcal{D}_i^2 \varepsilon_{(i)} \overset{\text{def}}{=} \ddot{\varepsilon}_{(i)} + \dot{\omega}_{(i)} \times \varepsilon_{(i)} + 2\omega_{(i)} \times \dot{\varepsilon}_{(i)} + \omega_{(i)} \times (\omega_{(i)} \times \varepsilon_{(i)})$$
$$= \mathcal{D}_i^2 d_{(i)} + u_{(1)}/M_1 - u_{(i)}/M_i, \tag{6.25}$$

where $\dot{\varepsilon}_{(i)}$ and $\ddot{\varepsilon}_{(i)}$ denote ${}^i d\varepsilon_{(i)}/dt$ and ${}^i d^2 \varepsilon_{(i)}/dt^2$ respectively.

**Proposition 6.1** *All solutions* $(\varepsilon_{(i)}, \dot{\varepsilon}_{(i)})(t)$ *of (6.25) with control law given by*

$$u_{(i)} = M_i(\mathrm{w}_{(i)} + \kappa_{2(i)}\dot{\varepsilon}_{(i)}), \quad i = 2, 3, \tag{6.26}$$

*with*

$$\mathrm{w}_{(i)} = u_{(1)}/M_1 + \mathbf{I}_i^{-1}(\omega_{(i)} \times (\mathbf{I}_i \omega_{(i)}) - \tau_{(i)}) \times \varepsilon_{(i)} - \langle \omega_{(i)}, \varepsilon_{(i)} \rangle \omega_{(i)}$$
$$+ (\kappa_{1(i)} + \|\omega_{(i)}\|^2)\varepsilon_{(i)} + \mathcal{D}_i^2 d_{(i)} \tag{6.27}$$

*tend to* $(0, 0) \in \mathbb{R}^6$ *as* $t \to \infty$ *for any real feedback gains* $\kappa_{1(i)}, \kappa_{2(i)} > 0, i = 2, 3$. *Moreover, the zero state of (6.25) with control law (6.26) is totally stable or stable under persistent disturbances [8].*

*Proof* Consider the following positive-definite function of $(\varepsilon_{(i)}, \dot{\varepsilon}_{(i)})$ defined on $\mathbb{R}^6$:

$$V_{1i} = (\kappa_{1(i)}\|\varepsilon_{(i)}\|^2 + \|\dot{\varepsilon}_{(i)}\|^2)/2, \quad \kappa_{1(i)} > 0. \tag{6.28}$$

The time rate-of-change of $V_{1i}$ along any solution of (6.25) is given by

$$dV_{1i}/dt = \langle \dot{\varepsilon}_{(i)}, (\mathcal{D}_{(i)}^2 d_{(i)} + u_{(1)}/M_1 - 2\omega_{(i)} \times \dot{\varepsilon}_{(i)} + \mathbf{I}_i^{-1}(\omega_{(i)} \times (\mathbf{I}_i\omega_{(i)}) - \tau_{(i)}) \times \varepsilon_{(i)}$$
$$- \omega_{(i)} \times (\omega_{(i)} \times \varepsilon_{(i)}) + \kappa_{1(i)}\varepsilon_{(i)} - u_{(i)}/M_i) \rangle$$
$$= \langle \dot{\varepsilon}_{(i)}, (\mathbf{w}_{(i)} - u_{(i)}/M_i) \rangle. \tag{6.29}$$

Using the vector identity: $\omega_{(i)} \times (\omega_{(i)} \times \varepsilon_{(i)}) = \omega_{(i)}\langle \omega_{(i)}, \varepsilon_{(i)} \rangle - \varepsilon_{(i)}\|\omega_{(i)}\|^2$, and setting $u_{(i)}$ in (6.29) to the control law (6.26) lead to

$$dV_{i1}/dt = -\kappa_{2(i)}\|\dot{\varepsilon}_{(i)}\|^2 \leq 0, \tag{6.30}$$

and (6.25) reduces to

$$\ddot{\varepsilon}_{(i)}(t) + \kappa_{2(i)}\dot{\varepsilon}_{(i)}(t) + 2\omega_{(i)}(t) \times \dot{\varepsilon}_{(i)}(t) + \kappa_{1(i)}\varepsilon_{(i)}(t) = 0. \tag{6.31}$$

Since $\langle \dot{\varepsilon}_{(i)}(t), (\omega_{(i)}(t) \times \dot{\varepsilon}_{(i)}(t)) \rangle = 0$ for all $\omega_{(i)}(t)$, it is evident that all solutions $(\varepsilon_{(i)}(t), \dot{\varepsilon}_{(i)}(t))$ of (6.31) tend to $(0, 0) \in \mathbb{R}^6$ as $t \to \infty$ for any $\kappa_{1(i)}, \kappa_{2(i)} > 0, i = 2, 3$. Moreover, the zero state of (6.31) is totally stable or stable under persistent disturbances [8]. This property implies that asymptotic stability of the zero state of (6.31) is preserved in the presence of small state-dependent perturbations. $\square$

We note that the realization of the control law (6.26) requires the knowledge of $u_{(1)}$ which must be communicated to the $i$th observer-spacecraft. Moreover, in real world situations, the components of the control $u_{(i)}$ are amplitude limited, i.e. $[u_{(i)}]^i = [u_{(i)1}, u_{(i)2}, u_{(i)3}]^{iT}$ satisfies $|u_{(i)j}| \leq F_{c(i)}, j = 1, 2, 3$, where $F_{c(i)}, i = 1, 2, 3$ are given positive constants. We consider again (6.29) rewritten in reference to a given orthonormal basis $B_i$ for $\mathcal{F}_i$:

$$dV_{1i}/dt = \sum_{j=1}^{3} \{[\dot{\varepsilon}_{(i)}]_j^i([\mathbf{w}_{(i)}]_j^i - u_{(i)j}/M_i)\}, \tag{6.32}$$

where $[\mathbf{w}_{(i)}]_j^i$ denotes the $j$th component of $[\mathbf{w}_{(i)}]^i$. If we set

$$u_{(i)j} = F_{c(i)}\text{sat}(g_{ij}(\omega_{(i)}, [\varepsilon_{(i)}]_j^i, [\dot{\varepsilon}_{(i)}]_j^i, u_{(1)}, d_{(i)})), \qquad j = 1, 2, 3, \tag{6.33}$$

where $\text{sat}(\cdot)$ denotes the saturation function defined by $\text{sat}(\alpha) = \text{sign}(\alpha)$ if $|\alpha| \geq 1$ and $\text{sat}(\alpha) = \alpha$ if $|\alpha| \leq 1$, and

$$g_{ij}(\omega_i, [\varepsilon_{(i)}]_j^i, [\dot{\varepsilon}_{(i)}]_j^i, u_{(1)}, d_{(i)}) \stackrel{\text{def}}{=} \frac{M_i}{F_{c(i)}}\left([\mathbf{w}_{(i)}]_j^i + \kappa_{2(i)}[\dot{\varepsilon}_{(i)}]_j^i\right), \qquad j = 1, 2, 3, \tag{6.34}$$

then

$$dV_{1i}/dt = \sum_{j=1}^{3} \{[\dot{\varepsilon}_{(i)}]_j^i ([w_{(i)}]_j^i - (F_{c(i)}/M_i)\text{sat}(g_{ij}(\omega_{(i)}, [\varepsilon_{(i)}]_j^i, [\dot{\varepsilon}_{(i)}]_j^i, u_{(1)}, d_{(i)})))\}.$$

(6.35)

Now, the tracking error equation (6.25) with control law (6.33) has the following representation with respect to basis $B_i$:

$$[\mathcal{D}_i^2 \varepsilon_{(i)}]_j^i = [\mathcal{D}_i^2 d_{(i)}) + u_{(1)}/M_1]_j^i - \frac{F_{c(i)}}{M_i}\text{sat}(g_{ij}(\omega_{(i)}, [\varepsilon_{(i)}]_j^i, [\dot{\varepsilon}_{(i)}]_j^i, u_{(1)}, d_{(i)})), \quad j = 1, 2, 3.$$

(6.36)

It can be readily verified that if $\|\omega_{(i)}(t)\|$, $\|u_{(1)}(t)\|$, $\|\mathcal{D}_i^2 d_{(i)}(t)\|$, and $\|\tau_{(i)}(t)\|$ are uniformly bounded for all $t \geq 0$, then the set

$$\{([\varepsilon_{(i)}]^i, [\dot{\varepsilon}_{(i)}]^i) \in \mathbb{R}^6 : |g_{ij}(\omega_{(i)}, [\varepsilon_{(i)}]_j^i, [\dot{\varepsilon}_{(i)}]_j^i, u_{(1)}, d_{(i)})| \leq 1, j = 1, 2, 3\}$$

(6.37)

contains a neighborhood of the zero state $(0, 0) \in \mathbb{R}^6$. Consequently, any solution $[(\varepsilon_{(i)}(t), \dot{\varepsilon}_{(i)}(t))]^i$ of (6.36) tends to the zero state as $t \to \infty$ for any real feedback gains $\kappa_{1(i)}$ and $\kappa_{2(i)} > 0$, and sufficiently small $\|([\varepsilon_{(i)}(0)]^i, [\dot{\varepsilon}_{(i)}(0)]^i)\|$.

### 6.2.1.3 Cluster-Formation Rotation and Attitude Synchronization

To clarify the notion of rotating the entire observer-spacecraft cluster-formation, we define the *cluster-formation body* $C_B(t)$ as the convex hull of $C(t)$ (i.e. the set of points formed by all convex combinations of the points in $C(t)$). Let b$(t)$ be a specified nonzero vector in $\mathbb{R}^3$. It is required to rotate $C_B(t)$ about an axis defined by the line $L(t) = a(t) + span\{b(t)\}$ with specified angular velocity $\Omega(t)$ as illustrated in Fig. 6.8, where a$(t)$ is a specified vector in $\mathbb{R}^3$. Moreover, the reference observer-spacecraft must rotate about a given body axis in *synchronism* with the cluster-formation body rotation, and the remaining observer-spacecraft must track the reference observer-spacecraft's attitude and angular velocity. This implies that the desired angular speed of the reference observer-spacecraft is $\|\Omega(t) + \Omega_o(t)\|$, and the desired angular velocity $\omega_{(i)}^d(t)$ and unit quaternion $q_{(i)}^d(t)$ for the $i$th observer-spacecraft, $i = 2, 3$, are $\omega_{(1)}(t)$ and $q_{(1)}(t)$ respectively, where $\Omega_o(t)$ is the angular velocity of $L(t)$ with respect to $\mathcal{F}_o$. Note that $\omega_{(i)}^d(t)$ and $q_{(i)}^d(t)$ must be *compatible* with each other in the sense that they satisfy

$$\frac{{}^o d\hat{q}_{(i)}^d}{dt} = (q_{(i)4}^d \omega_{(i)}^d - \omega_{(i)}^d \times \hat{q}_{(i)}^d)/2, \qquad \frac{{}^o dq_{(i)4}^d}{dt} = -\langle \omega_{(i)}^d, \hat{q}_{(i)}^d \rangle/2, \qquad (6.38)$$

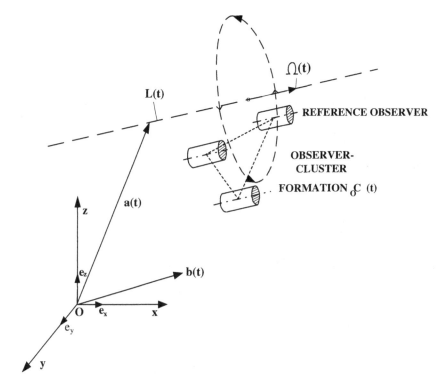

**Fig. 6.8** Rotation of the observer-spacecraft cluster about the line $L(t)$

and

$$\frac{^o d(\mathbf{I}_i \omega_{(i)}^d)}{dt} \stackrel{\text{def}}{=} \mathbf{I}_i \frac{^i d\omega_{(i)}^d}{dt} + \omega_{(i)}^d \times (\mathbf{I}_i \omega_{(i)}^d) = \tau_{(i)}^d, \tag{6.39}$$

where $q_{(i)}^d = [[\hat{q}_{(i)}^d]^T, q_{(i)4}^d]^T$, $\hat{q}_{(i)}^d = [q_{(i)1}^d, q_{(i)2}^d, q_{(i)3}^d]^T$.
To derive control laws for the reference and remaining observer-spacecraft to achieve attitude synchronization, we introduce

$$\delta q_{(i)} \stackrel{\text{def}}{=} q_{(i)}^d - q_{(i)} = [[\hat{q}_{(i)}^d - \hat{q}_{(i)}]^T, q_{i4}^d - q_{i4}]^T, \quad \delta\omega_{(i)} \stackrel{\text{def}}{=} \omega_{(i)}^d - \omega_{(i)}. \tag{6.40}$$

where $q_{(i)}^d = q_{(1)}$, and $\omega_{(i)}^d = \omega_{(1)}$, for $i = 2, 3$. It follows from (6.20), (6.21) to (6.40) that $\delta q_{(i)}$ and $\delta\omega_{(i)}$ satisfy

$$\frac{^o d\delta\hat{q}_{(i)}}{dt} = (q_{(i)4}^d \omega_{(i)}^d - q_{(i)4}\omega_{(i)} - \omega_{(i)}^d \times q_{(i)}^d + \omega_{(i)} \times \hat{q}_{(i)})/2, \tag{6.41}$$

$$\frac{^o d\delta q_{(1)4}}{dt} = -(\langle \omega_{(i)}^d, \hat{q}_{(i)}^d \rangle - \langle \omega_{(i)}, \hat{q}_{(i)} \rangle)/2. \tag{6.42}$$

and

$$\frac{^o d(\mathbf{I}_i \delta \omega_{(i)})}{dt} = \mathbf{I}_i \frac{^i d\delta \omega_{(i)}}{dt} + \delta \omega_{(i)} \times \mathbf{I}_i \omega_{(i)}^d + \omega_{(i)}^d \times \mathbf{I}_i \delta \omega_{(i)} - \delta \omega_{(i)} \times \mathbf{I}_i \delta \omega_{(i)} = \tau_{(i)}^d - \tau_{(i)}, \tag{6.43}$$

where $\tau_{(i)}^d$ is defined in (6.39).

**Proposition 6.2** *Given the desired attitude and angular velocity for the observer-spacecraft cluster-formation specified by $q_{(i)}^d(\cdot)$ and $\omega_{(i)}^d(\cdot), i = 2, 3$ respectively, under the action of the control law given by*

$$\tau_{(i)} = \kappa_{q(i)}\sigma + \kappa_{\omega(i)}\mathbf{I}_i \delta \omega_{(i)} + \tau_{(i)}^d - \omega_{(i)}^d \times (\mathbf{I}_i \delta \omega_{(i)})/2, \tag{6.44}$$

*where $\kappa_{q(i)}$ and $\kappa_{\omega(i)}$ are positive constant feedback gains, and*

$$\sigma = q_{(i)4}^d \delta \hat{q}_{(i)} - \delta q_{(i)4} \hat{q}_{(i)}^d - \hat{q}_{(i)}^d \times \delta \hat{q}_{(i)}, \tag{6.45}$$

*all solutions $(\delta \hat{q}_{(i)}(t), \delta q_{(1)4}(t), \delta \omega_{(i)}(t))$ of (6.41)–(6.43) tend to $(0, 0, 0) \in \mathbb{R}^7$ as $t \to \infty$.*

*Proof* Consider the following positive-definite function of $\delta \hat{q}_{(i)}, \delta q_{(i)4}$ and $\delta \omega_{(i)}$:

$$V_{2i} = \kappa_{q(i)}(\delta q_{(i)4}^2 + \langle \delta \hat{q}_{(i)}, \delta \hat{q}_{(i)} \rangle) + \langle \delta \omega_{(i)}, \mathbf{I}_i \delta \omega_{(i)} \rangle/2. \tag{6.46}$$

By requiring the time-rate-of-change of $V_{2i}$ along any solution of (6.41)–(6.43) to be non-positive for all $t \geq 0$, we obtain the attitude control law (6.44). It can be verified that under the action of the control law (6.44), any solution $[\delta \hat{q}_{(i)}(t), \delta q_{(i)4}(t), \delta \omega_{(i)}(t)]$ of (6.41)–(6.43) tends to $(0, 0, 0) \in \mathbb{R}^7$ as $t \to \infty$.  □

For the case where the control torques $\tau_{(i)}$ are amplitude limited (i.e. $[\tau_{(i)}]^i = [\tau_{(i)1}, \tau_{(i)2}, \tau_{(i)3}]^T$ satisfies $|\tau_{(i)j}| \leq T_{c(i)}, j = 1, 2, 3$), where $T_{c(i)}, i = 1, 2, 3$ are given positive constants, we set

$$[\tau_{(i)}]_j^i = T_{c(i)}\text{sat}\{[\kappa_{q(i)}\sigma + \kappa_{\omega(i)}\mathbf{I}_i \delta \omega_{(i)} + \tau_{(i)}^d - \omega_{(i)}^d \times (\mathbf{I}_i \delta \omega_{(i)})/2]_j^i/T_{c(i)}\}, \quad j = 1, 2, 3. \tag{6.47}$$

Here, since the control torques are amplitude limited, loss of synchronization can occur when the angular speed of formation rotation or the reference observer-spacecraft is so high that the remaining observer-spacecraft are unable to track the cluster-formation rotation. The possible occurrence of this phenomenon can be deduced from the following inequality obtained by first integrating (6.43) from time $t_o$ to $t$, and making use of the triangle inequality $\|\mathbf{a} - \mathbf{b}\| \geq \|\mathbf{a}\| - \|\mathbf{b}\|$ for norms of vectors $\mathbf{a}$ and $\mathbf{b}$:

$$\|\mathbf{I}_i(\delta\omega_{(i)}(t) - \delta\omega_{(i)}(t_o))\|_\infty = \left\| \int_{t_o}^t (\tau_{(i)}^d(s) - \tau_{(i)}(s))ds \right\|_\infty$$

$$\geq \left\| \int_{t_o}^t \tau_{(i)}^d(s)ds \right\|_\infty - \left\| \int_{t_o}^t \tau_{(i)}(s)ds \right\|_\infty \geq \left\| \int_{t_o}^t \tau_{(i)}^d(s)ds \right\|_\infty - \int_{t_o}^t \|\tau_{(i)}(s)\|_\infty ds$$

$$\geq \left\| \int_{t_o}^t \tau_{(i)}^d(s)ds \right\|_\infty - T_{c(i)}(t - t_o), \quad t \geq t_o, \tag{6.48}$$

where $\|\mathbf{a}\|_\infty \overset{\text{def}}{=} \max\{|a_i|; i = 1, 2, 3\}$ and $\mathbf{a} = (a_1, a_2, a_3)$. Now, suppose that the desired angular speed is such that the corresponding $\tau_{(i)}^d$'s satisfy $\| \int_{t_o}^t \tau_{(i)}^d(s)ds \|_\infty - T_{c(i)}(t - t_o) \geq \eta_i(t - t_o)$ for all $t$ in some time interval $I_{t_1} = [t_o, t_1]$, and some positive constant $\eta_i$. Then the norm of the deviation in the angular momentum of the $i$th observer-spacecraft from its desired value grows with time over $I_{t_1}$. Consequently, loss of synchronization results.

Now, we shall illustrate the cluster-formation rotation and attitude synchronization maneuvers for an observer-spacecraft triad moving in a triangular formation as shown in Fig. 6.9. Let $[r]^o = [x, y, z]^T$ denote the representation of a point $r$ with respect to an orthonormal basis $B_o = \{e_x, e_y, e_z\}$ in the inertial frame $\mathcal{F}_o$. We assume that the reference observer-spacecraft moves along a circular orbit $\mathcal{O} = \{(x, y, z) : x^2 + y^2 = a_o^2, z = z_o\}$ with radius $a_o$ and constant angular velocity $-\omega_o e_z$, and the position of the reference observer-spacecraft is given by

$$r_{(1)}(t) = (a_o \cos(\omega_o t))e_x - (a_o \sin(\omega_o t))e_y + z_o e_z. \tag{6.49}$$

To define the desired initial cluster-formation, we introduce constant deviation vectors $d_{(2)}^o$ and $d_{(3)}^o$:

$$d_{(2)}^o = d_{(2)x}^o e_x + d_{(2)y}^o e_y + d_{(2)z}^o e_z, \tag{6.50}$$

$$d_{(3)}^o = -d_{(2)x}^o e_x - d_{(2)y}^o e_y + d_{(2)z}^o e_z. \tag{6.51}$$

Then the desired initial cluster-formation pattern at time $t$ is specified by the point set

$$C^d(t) = \{r_{(1)}(t), r_{(1)}(t) + d_{(2)}^o, r_{(1)}(t) + d_{(3)}^o\} \tag{6.52}$$

and the cluster-formation body $C_B(t)$ is the plane domain bounded by the isosceles triangle with vertices given by $C^d(t)$. For a space interferometer, the combiner and collectors correspond to the reference and the remaining observer-spacecraft respectively. We assume that the observer-spacecraft body is a rigid cylinder with uniform mass density. Let $B_i = \{e_{x'}^i, e_{y'}^i, e_{z'}^i\}$ be an orthonormal basis for the body frame $\mathcal{F}_i$ for the $i$th observer-spacecraft such that $e_{z'}^i$ is along the cylinder axis which is also aligned with $e_z$. We consider two different cases: (a) the $x'$-axis is aligned with

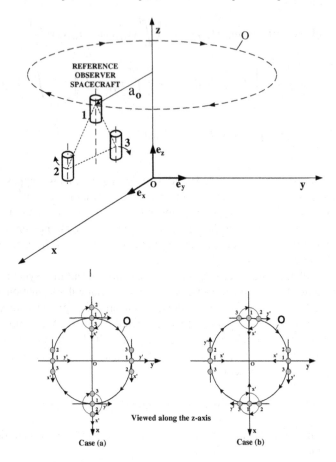

**Fig. 6.9** Rotation of an observer-spacecraft triad in a triangular formation

$x$-axis; and (b) the $x'$-axis passes through the center of the circular orbit $\mathcal{O}$ at all times, except during formation rotation (see Fig. 6.9). Case (a) corresponds to the situation where it is required to have the sensors or solar panels onboard the observer-spacecraft pointing to a fixed direction. Thus, the desired angular velocities $\omega_d^{(1)}$ for Cases (a) and (b) are zero and $-\omega_o e_z$ respectively.

Now, suppose that the objective is to rotate the triangular cluster-formation about the reference observer-spacecraft's cylindrical body axis with constant angular velocity $\Omega = -\omega_R e_z$. Thus, the rotation axis $L(t)$ for $C_B(t)$ is simply $span\{e_z\}$, and the deviation vectors $d_{(2)}(t)$ and $d_{(3)}(t)$ take on the form:

$$d_{(2)}(t) = (d^o_{(2)x} \cos(\omega_R t) - d^o_{(2)y} \sin(\omega_R t))e_x + (d^o_{(2)x} \sin(\omega_R t)$$
$$+ d^o_{(2)y} \cos(\omega_R t))e_y + d^o_{(2)z}e_z, \tag{6.53}$$

$$d_{(3)}(t) = (-d_{(2)x}^o \cos(\omega_R t) + d_{(2)y}^o \sin(\omega_R t))e_x$$
$$+(-d_{(2)x}^o \sin(\omega_R t) - d_{(2)y}^o \cos(\omega_R t))e_y + d_{(2)z}^o e_z. \qquad (6.54)$$

During formation rotation, the desired angular velocity $\omega_1^d$ for the reference observer-spacecraft relative to the inertial frame $\mathcal{F}_o$ is $-\omega_R e_z$ for Case (a), and $-(\omega_o + \omega_R)e_z$ for Case (b). The corresponding desired quaternion for the reference observer-spacecraft can be determined from (6.38) which can be written explicitly as

$$\frac{{}^o d}{dt} \begin{bmatrix} q_{(1)1}^d \\ q_{(1)2}^d \\ q_{(1)3}^d \\ q_{(1)4}^d \end{bmatrix} = \begin{bmatrix} 0 & \omega_T & 0 & 0 \\ -\omega_T & 0 & 0 & 0 \\ 0 & 0 & 0 & \omega_T \\ 0 & 0 & -\omega_T & 0 \end{bmatrix} \begin{bmatrix} q_{(1)1}^d \\ q_{(1)2}^d \\ q_{(1)3}^d \\ q_{(1)4}^d \end{bmatrix}, \qquad (6.55)$$

where $\omega_T = \omega_R/2$ for Case (a), and $\omega_T = (\omega_o + \omega_R)/2$ for Case (b).

The solution to (6.55) with initial condition $(\hat{q}_{(1)}(t_1)^T, q_{(1)4}(t_1))^T$ at time $t_1$ (the starting time for formation rotation) is given by

$$\begin{bmatrix} q_{(1)1}^d(t) \\ q_{(1)2}^d(t) \end{bmatrix} = \Phi(t - t_1) \begin{bmatrix} q_{(1)1}(t_1) \\ q_{(1)2}(t_1) \end{bmatrix}, \quad \begin{bmatrix} q_{(1)3}^d(t) \\ q_{(1)4}^d(t) \end{bmatrix} = \Phi(t - t_1) \begin{bmatrix} q_{(1)3}(t_1) \\ q_{(1)4}(t_1) \end{bmatrix}, \quad t \geq t_1,$$
$$(6.56)$$

where

$$\Phi(t - t_1) = \begin{bmatrix} \cos(\omega_T(t - t_1)) & \sin(\omega_T(t - t_1)) \\ -\sin(\omega_T(t - t_1)) & \cos(\omega_T(t - t_1)) \end{bmatrix}. \qquad (6.57)$$

**Simplification of Control Laws**

From (6.26) and (6.27), it is evident that the cluster formation-keeping control law consists of feedback terms involving $\varepsilon_{(i)}(t)$, $\dot{\varepsilon}_{(i)}(t)$, $\omega_{(i)}(t)$ and $\tau_{(i)}(t)$; and feedforward terms involving $\mathcal{D}_i^2 d_{(i)}(t)$ and the control thrust $u_{(1)}(t)$ of the reference observer-spacecraft. This control law is not amenable to physical realization. Therefore we shall consider a simplified version of control law (6.26) given by

$$u_{(i)}/M_i = \mathbf{K}_{1(i)}\varepsilon_{(i)} + \mathbf{K}_{2(i)}\dot{\varepsilon}_{(i)} + \tilde{w}_{(i)}, \qquad (6.58)$$

where $\mathbf{K}_{1(i)}$ and $\mathbf{K}_{2(i)}$ are positive self-adjoint linear transformations on $\mathbb{R}^3$ onto $\mathbb{R}^3$, and

$$\tilde{w}_{(i)} = u_{(1)}/M_1 + \mathcal{D}_i^2 d_{(i)}. \qquad (6.59)$$

Substituting (6.58) into (6.25) leads to the following linear time-varying differential equation for the tracking error:

$$\ddot{\varepsilon}_{(i)}(t) + \mathbf{R}_{(i)}(t)\dot{\varepsilon}_{(i)}(t) + \mathbf{S}_{(i)}(t)\varepsilon_{(i)}(t) = 0, \qquad (6.60)$$

where $\mathbf{R}_{(i)}(t)$ and $\mathbf{S}_{(i)}(t)$ are linear transformations on $\mathbb{R}^3$ into $\mathbb{R}^3$ defined by

$$\mathbf{R}_{(i)}(t) = \mathbf{K}_{2(i)} + 2\omega_i(t) \times (\cdot), \tag{6.61}$$

and

$$\mathbf{S}_{(i)}(t) = \dot{\omega}_{(i)}(t) \times (\cdot) + \omega_{(i)}(t)(\omega_{(i)}(t) \cdot) + (\mathbf{K}_{1(i)} - \|\omega_i(t)\|^2 \mathbf{I}). \tag{6.62}$$

Consider the following function defined on $\mathbb{R}^6$:

$$\tilde{V}_i = (\langle \varepsilon_{(i)}, \mathbf{K}_{1(i)} \varepsilon_{(i)} \rangle + \langle \dot{\varepsilon}_{(i)}, \dot{\varepsilon}_{(i)} \rangle + 2\gamma \langle \dot{\varepsilon}_{(i)}, \varepsilon_{(i)} \rangle)/2, \tag{6.63}$$

where $\gamma$ is a positive constant. Since $\mathbf{K}_{1(i)}$ is a positive self-adjoint linear transformation, hence

$$\langle \varepsilon_{(i)}, \mathbf{K}_{1(i)} \varepsilon_{(i)} \rangle \geq \lambda_{min}(\mathbf{K}_{1(i)}) \|\varepsilon_{(i)}\|^2, \tag{6.64}$$

where $\lambda_{min}(\mathbf{K}_{1(i)})$ denotes the minimum eigenvalue of $\mathbf{K}_{1(i)}$. From (6.64) and the inequality $\langle \dot{\varepsilon}_{(i)}, \varepsilon_{(i)} \rangle \geq -\|\dot{\varepsilon}_{(i)}\| \|\varepsilon_{(i)}\|$, it follows that $\tilde{V}_i$ satisfies the lower bound:

$$\tilde{V}_i \geq \frac{1}{2} \begin{bmatrix} \|\varepsilon_{(i)}\| \\ \|\dot{\varepsilon}_{(i)}\| \end{bmatrix}^T \mathbf{P} \begin{bmatrix} \|\varepsilon_{(i)}\| \\ \|\dot{\varepsilon}_{(i)}\| \end{bmatrix} \geq \frac{1}{2} \lambda_{min}(\mathbf{P}) \|(\varepsilon_{(i)}, \dot{\varepsilon}_{(i)})\|^2, \tag{6.65}$$

where $\|(\varepsilon_{(i)}, \dot{\varepsilon}_{(i)})\|^2 = \|\varepsilon_{(i)}\|^2 + \|\dot{\varepsilon}_{(i)}\|^2$, and

$$\mathbf{P} = \begin{bmatrix} \lambda_{min}(\mathbf{K}_{1(i)}) & -\gamma \\ -\gamma & 1 \end{bmatrix}. \tag{6.66}$$

Thus, if

$$\lambda_{min}(\mathbf{K}_{1(i)}) \geq \gamma^2, \tag{6.67}$$

then $\tilde{V}_i$ is positive definite on $\mathbb{R}^6$. We shall show that under certain mild conditions, the zero state of (6.60) is exponentially stable.

Consider the time-rate-of-change of $\tilde{V}_i$ given by

$$d\tilde{V}_i/dt = \gamma \langle \varepsilon_{(i)}, \|\omega_i\|^2 (\mathbf{I} - \mathbf{K}_{1(i)}) \varepsilon_{(i)} \rangle + \langle \dot{\varepsilon}_{(i)}, (\gamma \mathbf{I} - \mathbf{K}_{2(i)}) \dot{\varepsilon}_{(i)} \rangle - \gamma \langle \omega_{(i)}, \varepsilon_{(i)} \rangle^2$$
$$- \langle \dot{\varepsilon}_{(i)}, (\dot{\omega}_{(i)} \times \varepsilon_{(i)}) \rangle - \langle \omega_{(i)}, \varepsilon_{(i)} \rangle \langle \dot{\varepsilon}_{(i)}, \omega_{(i)} \rangle$$
$$+ \langle \varepsilon_{(i)}, (\|\omega_{(i)}\|^2 \mathbf{I} - \gamma \mathbf{K}_{2(i)}) \dot{\varepsilon}_{(i)} \rangle - 2\gamma \langle \varepsilon_{(i)}, (\omega_{(i)} \times \dot{\varepsilon}_{(i)}) \rangle. \tag{6.68}$$

From (6.21), it is evident that for bounded control torque $\tau_{(i)}$ (i.e. $\|\tau_{(i)}(t)\| \leq \hat{\tau}_{(i)} < \infty$ for all $t \geq 0$), there exist positive constants $\alpha_i$ and $\beta_i$ such that

$$\sup\{\|\omega_{(i)}(t)\|; t \geq 0\} \leq \alpha_i < \infty, \quad \sup\{\|\dot{\omega}_{(i)}(t)\|; t \geq 0\} \leq \beta_i < \infty. \tag{6.69}$$

Then $d\tilde{V}_i/dt$ satisfies the following estimate:

$$d\tilde{V}_i/dt \leq - \begin{bmatrix} \|\varepsilon_{(i)}\| \\ \|\dot{\varepsilon}_{(i)}\| \end{bmatrix}^T \mathbf{Q} \begin{bmatrix} \|\varepsilon_{(i)}\| \\ \|\dot{\varepsilon}_{(i)}\| \end{bmatrix} \qquad (6.70)$$

where

$$\mathbf{Q} = \begin{bmatrix} \gamma(\lambda_{min}(\mathbf{K}_{1(i)}) - \alpha_i^2) & \delta_{1i} \\ \delta_{1i} & \lambda_{min}(\mathbf{K}_{2(i)}) - \gamma \end{bmatrix}, \qquad (6.71)$$

and $\delta_{1i} = (\beta_i - 2\gamma\alpha_i + \alpha_i^2 + \gamma\|\mathbf{K}_{2(i)}\|)/2$. Now, if

$$\lambda_{min}(\mathbf{K}_{1(i)}) > \alpha_i^2, \quad \lambda_{min}(\mathbf{K}_{2(i)}) > \gamma, \qquad (6.72)$$

$$\gamma(\lambda_{min}(\mathbf{K}_{1(i)}) - \alpha_i^2)(\lambda_{min}(\mathbf{K}_{2(i)}) - \gamma) > \delta_{1i}^2, \qquad (6.73)$$

then $\mathbf{Q}$ is positive definite. Thus, under conditions (6.67), (6.72) and (6.73), we have

$$d\tilde{V}_i/dt \leq -\lambda_{min}(\mathbf{Q})(\|\varepsilon_{(i)}\|^2 + \|\dot{\varepsilon}_{(i)}\|^2)$$

$$\leq -2\{\lambda_{min}(\mathbf{Q})/\lambda_{min}(\mathbf{P})\}\tilde{V}_i. \qquad (6.74)$$

It follows that

$$\tilde{V}_i(t) \leq \tilde{V}_i(0) \exp\{-2(\lambda_{min}(\mathbf{Q})/\lambda_{min}(\mathbf{P}))t\} \qquad (6.75)$$

for all $t \geq 0$, which implies exponential stability of the zero state of (6.60).

*Remark 6.5*  The inclusion of the term $\tilde{w}_{(i)}$ in control law (6.58) is essential in achieving exponential stability. Since $\tilde{w}_{(i)}$ given by (6.59) involves the deviation vector $d_{(i)}(t)$ and control law $u_{c(1)}$ associated with the reference spacecraft, these data must be transmitted to the $i$th observer-spacecraft. If the $i$th spacecraft receives the deviation vector $d_{(i)}(t)$ and its velocity $\dot{d}_{(i)}(t)$ from the reference spacecraft, and has onboard sensors to measure the relative position $\rho_{i1}(t)$ and velocity $\dot{\rho}_{i1}(t)$, then the positional error and error rate can be determined by $\varepsilon_{(i)}(t) = d_{(i)}(t) - \rho_{i1}(t)$ and $\dot{\varepsilon}_{(i)}(t) = \dot{d}_{(i)}(t) - \dot{\rho}_{i1}(t)$ respectively. These data are required for the implementation of the simplified control law (6.58). A video of the computer simulation results depicting the performance of the proposed control laws for synchronized cluster-formation rotation is available [9].

### 6.2.1.4  Cluster-Formation Pattern Reconfiguration

In the observation of three-dimensional objects such as the case illustrated in Fig. 6.3, it may be necessary to alter the observer-cluster formation pattern imposed by the shape of the object under observation. Also, in any observer-cluster, failure in one or more observer units may occur, it is necessary to consider different options in maintaining the observer-cluster formation without impairing its function. The failure may take on various forms. When the failure is sufficiently severe such that the observer is no longer useful, its removal from the cluster is necessary. A possible option is to reconfigure the cluster-formation. Also, a change of observation objectives may require cluster-formation reconfiguration. For example, in the case of the interferometer, cluster-formation reconfiguration may be required to view a new target. In optimal cluster-formation reconfiguration, one may define a cost functional associated with the maneuvers of each spacecraft. It is required to determine the maneuver for each observer-spacecraft such that the total cost corresponding to the sum of the cost functionals of all the observer-spacecraft takes on its minimum value during cluster-formation reconfiguration.

In general, an observer-cluster may be composed of identical observer units as in the ESA/NASA Cluster [10], or not all identical spacecraft as in the proposed NASA DS-3 interferometer [11, 12]. In the latter case, the combiner and collector spacecraft have different structures and characteristics. Here, we consider an observer-cluster formation consisting of $P$ subsets of observers. Each subset $S_j \subset \mathcal{I}$ consists of $N_j$ *identical* observers and $\sum_{j=1}^{P} N_j = N$, where $\mathcal{I}$ denotes the index set $\{1, \dots, N\}$.

As before, let the actual and desired positions of the mass center of the $i$th observer-spacecraft at time t relative to $\mathcal{F}_o$ are denoted by $r_{(i)}(t)$ and $r_{(i)}^d(t)$ respectively. The point set $C(t) = \{r_{(1)}(t), \dots, r_{(N)}(t)\}$ generates a *cluster-formation pattern* at time $t$. The convex polytope $C_B(t)$ defined by $co(C(t))$ (the convex hull of $C(t)$) is referred to as the *cluster-formation body* at time $t$. In formation reconfiguration, a *desired cluster-formation pattern* given by $P^d(t) = \{p_{(1)}^d(t), \dots, p_{(N)}^d(t)\}$ defined for each $t$ in some time-interval $I_{t_f}$ is specified, where $p_{(i)}^d(t)$ corresponds to the position of the $i$th point in the desired cluster-formation pattern at time $t$ relative to $\mathcal{F}_o$. The cluster-formation pattern $P^d(t)$ or $C(t)$ is said to be *shape invariant* over some time interval $I_{t_f}$, if the Euclidean distance between any pair of distinct points in the cluster-formation pattern at time $t$ is constant for all $t \in I_{t_f}$. This implies that the geometric shape of the cluster-formation body does not vary with time over $I_{t_f}$.

In many physical situations involving subsets $S_j$ of identical spacecraft, it is only required that each point in the desired formation point set $C^d(t)$ be occupied by some spacecraft from a specified $S_j$. Thus, the $i$th element $p_{(i)}^d(t)$ in $P^d(t)$ may not correspond to $r_{(i)}^d(t)$. For example, for a spacecraft-triad in a triangular formation with $S_1 = \{1\}$, $S_2 = \{2, 3\}$, and $P^d(t) = \{p_{(1)}^d(t), p_{(2)}^d(t), p_{(3)}^d(t)\}$, we only require $p_{(1)}^d(t) = r_{(1)}^d(t)$, $(p_{(2)}^d(t), p_{(3)}^d(t)) = (r_{(2)}^d(t), r_{(3)}^d(t))$ or $(r_{(3)}^d(t), r_{(2)}^d(t))$.

In general, cluster-formation reconfiguration can be classified into two basic types. In Type 1, each observer-spacecraft is required to occupy a specified position in the desired reconfigured formation, whereas in Type 2, a specified position in the desired reconfigured formation may be occupied by *any* observer-spacecraft of a particular type. The total number of observer-spacecraft before and after reconfiguration may differ due to the presence of failed spacecraft, and/or augmentation of the number of observer-spacecraft to attain a larger cluster-formation. This situation can also arise when a large cluster-formation is partitioned into a number of small formations, or enlarged by merging a collection of small formations.

In the case where the total number of observer-spacecraft is less than that of the initial formation, it is possible that only a limited number of useful formation patterns can be generated from the reduced number of spacecraft, especially when the spacecraft are not all identical. In fact, it may be necessary to remove additional spacecraft before a useful geometric formation pattern can be obtained. For example, consider an initial formation consisting of two different types of equally numbered observer-spacecraft placed on a circle with equal spacing. In order to retain the circle formation with equal spacecraft spacing after reconfiguration, the number of these two types of spacecraft must be the same. Thus, any removal must be made in pairs of nonidentical spacecraft. In what follows, we assume that the total number of observer-spacecraft before initiation of the reconfiguration process has been adjusted to coincide with that in the desired reconfigured formation.

In optimal cluster-formation reconfiguration, it is required to move an observer-spacecraft from its position in the formation pattern before reconfiguration to a specified position in the desired formation pattern such that a given cost functional takes on its minimum value over all admissible maneuvers. When a spacecraft is under the influence of a gravitational field, its optimal trajectory is generally a space curve in $\mathbb{R}^3$. In the absence of gravitational field and disturbance forces, the optimal formation reconfiguration problem can be simplified by requiring the spacecraft to follow a straightline path connecting its initial and desired positions with minimum cost. These optimal maneuvers associated with individual observer-spacecraft will be referred to hereafter as *basic optimal maneuvers*. The optimal cluster-formation reconfiguration problem is to find the basic optimal maneuver for each spacecraft such that the total cost corresponding to the sum of the optimal cost associated with each observer-spacecraft is minimized. In what follows, we shall consider first the basic optimal maneuvers. Special attention is focused on the important problem where the cost corresponds to fuel expenditure.

### (i) Basic Optimal Maneuvers

Let the cluster-formation pattern without reconfiguration at $t \geq 0$ be denoted by $C^o(t) = \{r^o_{(1)}(t), \ldots, r^o_{(N)}(t)\}$, where $r^o_{(i)}(t)$ denotes the position of the $i$th observer spacecraft in $\mathcal{F}_o$ at time $t$. The evolution of $r^o_{(i)}(t)$ with time $t$ is described by

$$\frac{d^2 r^o_{(i)}(t)}{dt^2} = \frac{u^o_{c(i)}(t) + f_{g(i)}(t, r^o_{(i)}(t))}{M_i}, \quad i = 1, \ldots, N, \qquad (6.76)$$

where $u^o_{c(i)} = u^o_{c(i)}(t)$ is a given control thrust program for generating $r^o_{(i)} = r^o_{(i)}(t)$; $f_{g(i)} = f_{gi}(t, r^o_{(i)}(t))$ is the gravitational and/or disturbance force; and $M_i$ is the mass of the $i$th spacecraft.

Let the actual and desired positions $r_{(i)}(t)$ and $r^d_{(i)}(t)$ of the $i$th observer-spacecraft cluster during formation reconfiguration be described respectively by

$$\frac{d^2 r^d_{(i)}(t)}{dt^2} = \frac{u^d_{c(i)}(t) + f_{g(i)}(t, r^d_{(i)}(t))}{M_i}, \quad \frac{d^2 r_{(i)}(t)}{dt^2} = \frac{u_{c(i)}(t) + f_{g(i)}(t, r_{(i)}(t))}{M_i}, \tag{6.77}$$

where $u_{c(i)} = u_{c(i)}(t)$ denotes the actual control thrust for the $i$th observer-spacecraft, and $u^d_{(i)} = u^d_{(i)}(t)$ is the required thrust program for generating $r^d_{(i)} = r^d_{(i)}(t)$. We assume that the admissible control thrusts $u_{c(i)} = u_{c(i)}(t)$ are piecewise continuous functions of $t$ defined on some time interval $[0, t_i]$ taking their values in a given compact subset $\Omega_{(i)}$ of $\mathbb{R}^3$. A basic optimal maneuver problem is to find an admissible control thrust program which steers the $i$th observer-spacecraft from its initial state $(r_{(i)}(0), \dot{r}_{(i)}(0))$ at cluster-formation reconfiguration starting at time $t = 0$ to its target state $(r^d_{(i)}(t_i), \dot{r}^d_{(i)}(t_i))$ at some time $t_i > 0$ (specified or free) such that a given cost functional $\mathbb{C}_i = \mathbb{C}_i(r_{(i)}(\cdot), \dot{r}_{(i)}(\cdot), u_{c(i)}(\cdot))$ takes on its minimum value. This class of problems for single spacecraft with various cost functionals has been studied extensively by Lawden [13], Marec [14, 15], and Carter [16] (see also surveys by Bell [17] and Robinson [18]). Almost all these results are in the form of optimal open-loop control thrust programs. In the general optimal cluster-formation reconfiguration problem, we assume that the basic optimal maneuver $r^*_{(i)} = r^*_{(i)}(t)$ for each individual spacecraft with any given target state $(r^d_{(i)}(t_i), \dot{r}^d_{(i)}(t_i))$ along with its corresponding optimal cost $\mathbb{C}_i(r^*_{(i)}(\cdot), \dot{r}^*_{(i)}(\cdot), u^*_{c(i)}(\cdot))$ have been be determined.

For the special case of a free-flying spacecraft in the absence of gravitational field and disturbance forces (i.e. $f_{g(i)} = 0$), we let

$$\rho^o_{d(i)}(t) = r^d_{(i)}(t) - r^o_{(i)}(t), \quad \Delta_{(i)}(t) = \rho^o_{d(i)}(t)/\|\rho^o_{d(i)}(t)\|, \tag{6.78}$$

where $\|\cdot\|$ denotes the usual Euclidean norm. A basic optimal maneuver is to translate the $i$th spacecraft in the direction $\Delta_{(i)}(t)$ to its target position $r^d_{(i)}(t_i)$ at some time $t_i > 0$ with minimum fuel expenditure.

Consider the projection of $(r^d_{(i)}(t) - r_{(i)}(t))$ onto the direction $\Delta_{(i)}(t)$ given by

$$s_{(i)}(t) = \langle r^d_{(i)}(t) - r_{(i)}(t), \Delta_{(i)}(t)\rangle, \tag{6.79}$$

where $\langle \cdot, \cdot \rangle$ denotes the usual inner product on $\mathbb{R}^3$. By differentiation and making use of (6.77) and (6.78), we have

$$\frac{d^2 s_{(i)}(t)}{dt^2} = \left\langle \frac{u_{(i)}^d(t) - u_{c(i)}(t)}{M_i}, \Delta_{(i)}(t) \right\rangle + 2 \left\langle \frac{d(r_{(i)}^d(t) - r_{(i)}(t))}{dt}, \frac{d\Delta_{(i)}(t)}{dt} \right\rangle$$

$$+ \left\langle r_{(i)}^d(t) - r_{(i)}(t), \frac{d^2 \Delta_{(i)}(t)}{dt^2} \right\rangle. \tag{6.80}$$

We assume that $r_{(i)}^o(t)$ and $r_{(i)}^d(t)$ have the forms

$$r_{(i)}^o(t) = r_{(i)}^o(0) + \xi(t), \qquad r_{(i)}^d(t) = r_{(i)}^d(0) + \xi(t), \tag{6.81}$$

where $\xi = \xi(t)$ is a given function of $t$, representing the formation drift vector. Then, $\Delta_{(i)}(t) = (r_{(i)}^d(0) - r_{(i)}^o(0))/\|r_{(i)}^d(0) - r_{(i)}^o(0)\|$, a constant vector, implying that the distance between $r_{(i)}^d(t)$ and $r_{(i)}^o(t)$ is time-invariant, and $u_{c(i)}^d(t) = u_{c(i)}^o(t)$. Thus, (6.80) reduces to

$$\frac{d^2 s_{(i)}(t)}{dt^2} = \left\langle \frac{u_{(i)}^d(t) - u_{c(i)}(t)}{M_i}, \Delta_{(i)}(t) \right\rangle \overset{\text{def}}{=} \hat{u}_{(i)}(t)/M_i, \tag{6.82}$$

where $\hat{u}_{(i)}$ corresponds to the effective control along the direction $\Delta_{(i)}(t)$. We assume that the admissible $\hat{u}_{(i)} = \hat{u}_{(i)}(t)$ are piecewise continuous functions of $t$ satisfying $|\hat{u}_{(i)}(t)| \le \bar{u}_{(i)}$ for all $t$, where $\bar{u}_{(i)}$ is a specified positive constant. Let $\tilde{s}_{(i)} = s_{(i)} M_i / \bar{u}_{(i)}$ and $u_{(i)} = \hat{u}_{(i)}/\bar{u}_{(i)}$. Then, (6.82) can be rewritten in the following normalized form:

$$\frac{d^2 \tilde{s}_{(i)}}{dt^2} = u_{(i)}. \tag{6.83}$$

Now, given a terminal time $t_i < \infty$, the minimum-fuel translational manuever problem is to find an admissible normalized control $u_{(i)}^* = u_{(i)}^*(t)$ defined on the time interval $[0, t_i]$ which steers the initial state $(\tilde{s}_{(i)}, d\tilde{s}_{(i)}/dt)(0) = (\|\rho_{di}^o(0)\| M_i / \bar{u}_{(i)}, 0)$ to the target state $(\tilde{s}_{(i)}, d\tilde{s}_{(i)}/dt)(t_i) = (0, 0)$ at time $t_i$ such that the fuel expenditure associated with the translational maneuver given by

$$F_i(u_{(i)}) = \alpha_i \int_0^{t_i} |u_{(i)}(t)| dt \tag{6.84}$$

is minimized, where $\alpha_i$ is a specified positive proportionality constant.

The complete solution to this problem is given in [18]. Let $x_1 = \tilde{s}_{(i)}$, and $x_2 = dx_1/dt$ so that (6.83) takes on the standard form:

$$\frac{d}{dt} \begin{bmatrix} x_1 \\ x_2 \end{bmatrix} = \begin{bmatrix} x_2 \\ u_{(i)} \end{bmatrix}, \tag{6.85}$$

**Fig. 6.10** The sets $R_i$, $i = 1, \ldots, 4$ and a minimum-fuel trajectory

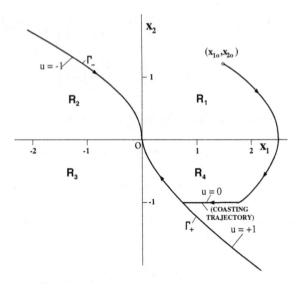

and the optimal control problem corresponds to finding an admissible control defined on the time interval $[0, t_i]$ such that the initial state $x_o = (x_{1o}, x_{2o})^T$ is steered to the zero state at time $t_i$ with minimum fuel expenditure.

Let

$$\Gamma_+ = \{(x_1, x_2) \in \mathbb{R}_2 : x_1 = \frac{1}{2}x_2^2, x_2 \leq 0\}; \quad \Gamma_- = \{(x_1, x_2) \in \mathbb{R}_2 : -\frac{1}{2}x_2^2, x_2 \geq 0\}; \tag{6.86}$$

$$R_1 = \{(x_1, x_2) \in \mathbb{R}_2 : x_2 \geq 0; x_1 > x_1', \text{, where } (x_1', x_2) \in \Gamma_-\};$$
$$R_2 = \{(x_1, x_2) \in \mathbb{R}_2 : x_2 > 0; x_1 < x_1', \text{where } (x_1', x_2) \in \Gamma_-\};$$
$$R_3 = \{(x_1, x_2) \in \mathbb{R}_2 : x_2 \leq 0; x_1 < x_1', \text{where } (x_1', x_2) \in \Gamma_+\};$$
$$R_4 = \{(x_1, x_2) \in \mathbb{R}_2 : x_2 < 0; x_1 > x_1', \text{where } (x_1', x_2) \in \Gamma_+\}. \tag{6.87}$$

The above sets $R_i$, $i = 1, \ldots, 4$ are shown in Fig. 6.10.

Given an initial state $\mathbf{x}_o = (x_{1o}, x_{2o})^T \in \mathbb{R}_2$ at time $t = 0$, the foregoing problem has a solution if and only if $t_i \geq t_i^*$, where $t_i^*$ is the minimum time for steering $\mathbf{x}_o$ to the zero state, or

$$t_i \geq t_i^* = \begin{cases} x_{2o} + \sqrt{4x_{1o} + 2x_{2o}^2} & \text{for } \mathbf{x}_o \in R_1 \cup R_4, \\ -x_{2o} + \sqrt{-4x_{1o} + 2x_{2o}^2} & \text{for } \mathbf{x}_o \in R_2 \cup R_3, \\ |x_{2o}| & \text{for } \mathbf{x}_o \in \Gamma_+ \cup \Gamma_-. \end{cases} \tag{6.88}$$

Assuming that condition (6.88) is satisfied, then for all $x_o \in R_1$ and all $0 < t_i < \infty$; or $x_o \in R_4$ and all $t_i \leq -\frac{1}{2}x_{2o} - \frac{x_{1o}}{x_{2o}}$, the optimal control $u^*_{(i)} = u^*_{(i))}(t)$ is given uniquely by

$$u^*_{(i)}(t) = \begin{cases} -1 & \text{for } 0 \leq t < t_{i1}, \\ 0 & \text{for } t_{i1} \leq t < t_{i2}, \\ 1 & \text{for } t_{i2} \leq t \leq t_i. \end{cases} \tag{6.89}$$

where

$$t_{i1} = \frac{1}{2}\left\{t_i + x_{2o} - \sqrt{(t_i - x_{2o})^2 - 4x_{1o} - 2x_{2o}^2}\right\},$$

$$t_{i2} = \frac{1}{2}\left\{t_i + x_{2o} + \sqrt{(t_i - x_{2o})^2 - 4x_{1o} - 2x_{2o}^2}\right\}. \tag{6.90}$$

The corresponding minimum fuel expenditure is given by

$$\mathbb{C}_i(u_i^*) = \alpha_i \left\{t_i - \sqrt{(t_i - x_{2o})^2 - 4x_{1o} + 2x_{2o}^2}\right\}. \tag{6.91}$$

A realization of the foregoing optimal control in feedback form is given by

$$u^*_{(i)}(x_1, x_2) = -\text{Sgn}(x_1 + \frac{1}{2}x_2|x_2|) - \text{Sgn}(x_1 + m_\beta x_2|x_2|), \tag{6.92}$$

where

$$m_\beta = \frac{\beta}{2\beta - 2\sqrt{\beta(\beta - 1)} - 1} - \frac{1}{2}, \quad \beta = t_i/t_i^*. \tag{6.93}$$

The foregoing optimal control is relevant to the minimum-fuel formation reconfiguration problem to be considered later. The optimal controls corresponding to other conditions on $x_o$ and $t_i$ are given in [18]. It can be seen from (6.89) that an optimal trajectory consists of the trajectory due to an initial full thrust in the direction of $\Delta_{(i)}(t)$, a coasting trajectory, and a trajectory due to full thrust opposite the direction of $\Delta_{(i)}(t)$ (see Fig. 6.10).

Another basic optimal maneuver is to transpose or interchange the positions of an observer-spacecraft pair, say the $i$th and $j$th observer-spacecraft, with minimum cost. In this case, we set

$$r^d_{(i)}(t) = r^o_{(j)}(t), \quad r^d_{(j)}(t) = r^o_{(i)}(t). \tag{6.94}$$

In general, the optimal trajectories in world space for the $i$th and $j$th observer-spacecraft are distinct curves with unequal optimal costs.

For the case of a free-flying observer-spacecraft in the absence of gravitational field and disturbance forces, we define

$$\rho^o_{d(i)}(t) = r^o_{(j)}(t) - r^o_{(i)}(t), \quad \Delta_{(i)}(t) = \rho^o_{d(i)}(t)/\|\rho^o_{d(i)}(t)\|, \quad \Delta_{(j)}(t) = -\Delta_{(i)}(t).$$
$$(6.95)$$

As before, we consider the projection of $(r_{(i)}(t) - r^o_{(i)}(t))$ onto the direction $\Delta_{(i)}(t)$ given by (6.95). Under the assumption that

$$r^o_{(i)}(t) = r^o_{(i)}(0) + \xi(t), \quad r^o_{(j)}(t) = r^o_{(j)}(0) + \xi(t), \tag{6.96}$$

where $\xi = \xi(t)$ is the formation drift vector as in (6.81), then $\Delta_{(i)}(t) = r^o_{(j)}(0) - r^o_{(i)}(0)$ is a constant vector, implying that the initial or desired formation patterns for $t \geq 0$ are shape-invariant. Thus, (6.82) for $s_{(i)}(t)$ is valid here. Consequently, the results mentioned earlier for the minimum-fuel translational maneuver are also applicable to this case.

Note that requiring the $i$th and $j$th observer-spacecraft to move along their respective directions $\Delta_{(i)}(t)$ and $\Delta_{(j)}(t) = -\Delta_{(i)}(t)$ simultaneously during transposition could result in a collision. This situation can be avoided by executing a collision-avoidance maneuver when the observer-spacecraft are close to each other, or by introducing a side-stepping motion before initiating the transposition, and a recovery motion before terminating the transposition. The abovementioned situation can also be avoided by moving more than two spacecraft simultaneously along straightline paths to their desired positions without transposition.

### (i) Type 1 Formation Reconfiguration

We assume that the formation patterns $C^o(t) = \{r^o_{(1)}(t), \ldots, r^o_{(N)}(t)\}, t \geq 0$, before reconfiguration are shape-invariant. Let the desired cluster-formation pattern at time $t$ be denoted by $P^d(t) = \{p^d_{(1)}(t), \ldots, p^d_{(N)}(t)\}$. We further assume that there is a one-to-one correspondence between the elements of $P^d(t)$ and the set of desired observer-spacecraft positions $\{r^d_{(1)}(t), \ldots, r^d_{(N)}\}$ for $i, j, \in \mathcal{I}$.

First, we consider the simplest case where $r^o_{(i)}(t)$ and $r^d_{(i)}(t) = r^d_{(j)}(t)$ are related by a translation, i.e.

$$r^d_{(i)}(t) = a_{(i)} + r^o_{(i)}(t), \quad i \in \mathcal{I}, \tag{6.97}$$

where $a_{(i)}$ is a given constant vector in $\mathbb{R}_3$, and $\mathcal{I}$ denotes the observer-spacecraft index set. Evidently, $C^o(t)$, and $\{r^d_{(1)}(t), \ldots, r^d_{(N)}(t)\}, t \geq 0$, are also shape invariant. In this case, the formation reconfiguration can be achieved by moving one or a small group of observer-spacecraft at a time. Let $\mathbb{C}^*_i$ denote the optimal cost corresponding to the optimal basic maneuver $r^*_{(i)}(\cdot)$ steering $(r_{(i)}, \dot{r}_{(i)})(0)$ at $t = 0$ to $(r^d_{(i)}, \dot{r}^d_{(i)})(t_i)$ at some time $t_i$. Then, the total optimal cost is $\mathbb{C}^*_T = \sum^N_{i=1} \mathbb{C}^*_i$.

In the case of free-flying observer-spacecraft, we assume that $r^d_{(i)}(t)$ is *visible* from $r^o_{(i)}(t)$ at any time $t \geq 0$ for any $i \in \mathcal{I}$, (i.e. the line segment joining $r^d_{(i)}(t)$ and $r^o_{(i0)}(t)$ does not pass through $r^o_{(j)}(t)$ or $r^d_{(j)}(t)$ for any $j \in \mathcal{I} - \{i\}$), then, given a transfer time $t_i \geq t^*_i$, the $i$th observer-spacecraft can be steered from $r^o_{(i)}(0)$ to $r^d_{(i)}(t_i)$ with minimum fuel expenditure by a control of the form (6.89). The formation reconfiguration can be achieved by moving one or a small group of observer-spacecraft at a time. Thus, the minimum total fuel expenditure for formation reconfiguration is given by $\mathbb{C}_T = \sum_{i=1}^{N} \mathbb{C}_i(u_i^*)$. Note that given an upper bound $\bar{\mathbb{C}}_T$ for $\mathbb{C}_T$, we can always choose a set of transfer times $t_1, \ldots, t_N$ satisfying $t_i \geq t_i^*, i = 1, \ldots, N$ such that $\mathbb{C}_T \leq \bar{\mathbb{C}}_T$.

Next, we consider the case where $r_{(i)}^d(t) = r_{(j)}^o(t)$ for some $j \in \mathcal{I} - \{i\}$. Under the assumption that the initial cluster-formation patterns are shape-invariant, $(r_{(i)}^d(t) - r_{(i)}^o(t))$ or $(r_{(j)}^o(t) - r_{(i)}^o(t))$ is a constant nonzero vector for all $t \geq 0$. Again, we assume that $r_{(j)}^o(t)$ is visible from $r_{(i)}^o(t)$ at any time $t \geq 0$. Although it is possible to reconfigure the cluster-formation by moving all the observer-spacecraft to their desired positions simultaneously, such an approach may result in chaos due to control system malfunctions and/or possible collisions between the observer-spacecraft. Therefore we propose a more prudent approach to reconfigure the formation by introducing a sequence of maneuvers involving a small number of spacecraft at a time. To clarify ideas, we first consider in detail the case involving transpositions between spacecraft pairs. Thus, the optimal formation reconfiguration problem is to find a sequence of transpositions with minimum total cost.

Let $\sigma$ be a permutation of the index set $\mathcal{I}$ (i.e. a one-to-one mapping of $\mathcal{I}$ onto $\mathcal{I}$ such that for $i \in \mathcal{I}, \sigma(i) = s_i \in \mathcal{I}$; and $\sigma(i) = \sigma(j)$ if and only if $i = j$) described by the usual notation

$$\sigma = \begin{pmatrix} 1 \ldots N \\ s_1 \ldots s_N \end{pmatrix}. \tag{6.98}$$

Let $\mathcal{G}(\mathcal{I})$ denote the set of all permutations of $\mathcal{I}$. For $\sigma, \sigma' \in \mathcal{S}(\mathcal{I})$, we define their product $\sigma\sigma'$ by $(\sigma\sigma')(i) = \sigma(\sigma'(i)), i \in \mathcal{I}$; the identity by $\mathbf{i}(i) = i$ for each $i \in \mathcal{I}$; and $\sigma^{-1}$ (the inverse of $\sigma$) by $\sigma^{-1}\sigma = \sigma\sigma^{-1} = \mathbf{i}$. It can be readily verified that $\mathcal{G}(\mathcal{I})$ is a group.

Let the initial spacecraft formation be represented by a permutation of $\mathcal{I}$ denoted by

$$\sigma_o = \begin{pmatrix} 1 \ldots N \\ s_{o1} \ldots s_{oN} \end{pmatrix}. \tag{6.99}$$

Since the formation body is assumed to be shape-invariant under reconfiguration, the desired formations correspond to other permutations of $\mathcal{I}$. The set of all formations corresponding to permutations of $\sigma_o$ forms a group with $N!$ elements.

Now, consider a *transposition* defined by the permutation: $\tau(i') = j'$ and $\tau(j') = i'$ for $i' \neq j' \in \mathcal{I}$, and $\tau(i) = i$ whenever $i \neq i'$ and $i \neq j'$. The following property can be verified by induction:

**Property A** *Every permutation $\sigma \in \mathcal{G}(\mathcal{I})$ is a product of transpositions.*

The above property implies that any desired reconfigured formation is attainable from any initial formation by some product of transpositions. To facilitate the computations, we use the column vector $s = [s_1, \ldots, s_N]^T$ ($s_i \in \mathcal{I}$, and $s_i = s_j$ if and

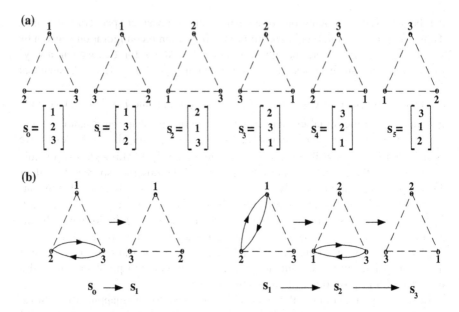

**Fig. 6.11** Triangular formation patterns and their reconfigurations attained by transpositions. **a** All possible (3!) *triangular* formations. **b** Formations attained by transpositions

only if $i = j$) to denote a formation or a permutation of $\mathcal{I}$. The transformation that takes $s = [s_1, \ldots, s_N]^T$ into $s' = [s_{1'}, \ldots, s_{N'}]^T$ can be represented by the $N \times N$ permutation matrix $\mathbb{P}_{(s',s)}$ in which the elements in the $i$th column (for each i) are all zero except for the one in the $i'$th row, which is unity. Thus, $s' = \mathbb{P}_{(s',s)}\,s$. Evidently, any permutation group is isomorphic with the group of corresponding permutation matrices. The transposition defined earlier can be represented by a special $N \times N$ permutation matrix $\mathbb{T}_{(i',j')}, i' \neq j' \in \mathcal{I}$, in which the $(i', j')$th and $(j', i')$th elements and the diagonal elements $(i \neq i', j')$ are unity, and the remaining elements are all zero.

Now, given an initial formation $s_o$ and a desired formation $s_d$, there exists a permutation matrix $\mathbb{P}_{(s_d,s_o)}$ represented by a product of transpositions $\mathbb{T}_{(i,j)}, \ldots, \mathbb{T}_{(i',j')}$ such that $s_d = \mathbb{T}_{(i,j)} \circ \ldots \circ \mathbb{T}_{(i',j')}s_o$. In general, the product of transpositions that take $s_o$ to $s_d$ is nonunique. Moreover, there may exist transpositions that are not admissible due to physical constraints derived from formation geometry and other considerations.

To clarify the foregoing notions, we consider a simple example.

*Example 6.1* Consider an observer-triad cluster in triangular formations with index set $\mathcal{I} = \{1, 2, 3\}$. Let the initial formation (see Fig. 6.11) be represented by $s_o = [1, 2, 3]^T$. Then the remaining possible triangular formations generated by permuting $s_o$ have the representations: $s_1 = [1, 3, 2]^T$, $s_2 = [2, 1, 3]^T$, $s_3 = [2, 3, 1]^T$, $s_4 = [3, 2, 1]^T$, and $s_5 = [3, 1, 2]^T$. Evidently, $s_1$ can be generated from $s_o$ by applying the transposition represented by the matrix:

$$\mathbb{T}_{(2,3)} = \begin{bmatrix} 1 & 0 & 0 \\ 0 & 0 & 1 \\ 0 & 1 & 0 \end{bmatrix}. \tag{6.100}$$

Formation $s_5$ can be generated from $s_o$ by applying the permutation matrix:

$$\mathbb{P}_{(s_5,s_o)} = \begin{bmatrix} 0 & 0 & 1 \\ 1 & 0 & 0 \\ 0 & 1 & 0 \end{bmatrix}, \tag{6.101}$$

which has three distinct decompositions in the form of products of two transpositions given by:

$$\mathbb{P}_{(s_5,s_o)} = \mathbb{T}_{(3,1)}\mathbb{T}_{(1,2)} = \mathbb{T}_{(2,3)}\mathbb{T}_{(3,1)} = \mathbb{T}_{(1,2)}\mathbb{T}_{(2,3)}, \tag{6.102}$$

where

$$\mathbb{T}_{(3,1)} = \begin{bmatrix} 0 & 0 & 1 \\ 0 & 1 & 0 \\ 1 & 0 & 0 \end{bmatrix}, \quad \mathbb{T}_{(1,2)} = \begin{bmatrix} 0 & 1 & 0 \\ 1 & 0 & 0 \\ 0 & 0 & 1 \end{bmatrix}. \tag{6.103}$$

Now, suppose that the initial formation is a *line* formation represented by $s = [1, 2, 3]^T$ as before. The reconfigured formations are line formations generated by permuting $s_o$ as shown in Fig. 6.12. Clearly, these line formations have the same representations as in the triangular formation case. Suppose that a transposition between a point-observer pair is *admissible* only if they are mutually visible. Then, the transposition $\mathbb{T}_{(3,1)}$ is not admissible. Consequently, $\mathbb{P}_{(s_5,s_o)}$ has a unique admissible decomposition $\mathbb{T}_{(1,2)}\mathbb{T}_{(2,3)}$.

To solve the optimal formation reconfiguration problem, we propose the following basic steps:

**Step 1** Construct the $N \times N$ optimal cost matrices $\mathbb{C}^{(n)}, n = 1, \ldots, N$, whose $(i, j)$th element $\mathbb{C}_{ij}^{(n)}$ corresponds to the optimal cost associated with moving the $n$th observer-spacecraft from the $j$th to the $i$th location in the formation. In general, in the presence of gravitational field and/or disturbance forces, $\mathbb{C}^{(n)}$ is a non-symmetric matrix with zero diagonal elements. For the minimum fuel case, the numerical values for $\mathbb{C}_{ij}^{(n)}$ can be computed using (6.91). For a transposition maneuver involving moving the $n$th spacecraft from the $j$th to the $i$th position in the formation, and the $m$th observer-spacecraft from the $i$th to the $j$th position in the formation, the total cost for transposition is $\mathbb{C}_{ij}^{(n)} + \mathbb{C}_{ji}^{(m)}$.

**Step 2** Given the initial and desired formations represented by $s_o$ and $s_d$ respectively, obtain the permutation matrix $\mathbb{P}_{(s_d,s_o)}$ and find all its possible decompositions in the form of products of admissible transpositions.

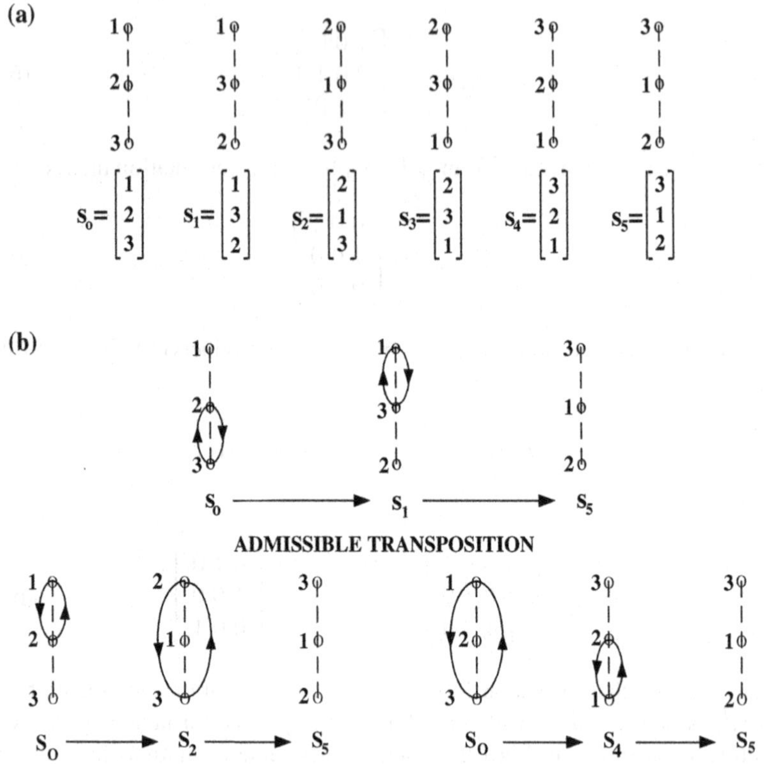

**(a)**

**(b)**

ADMISSIBLE TRANSPOSITION

**Fig. 6.12** Line formations for an observer-triad cluster. **a** All possible (3!) line formations. **b, c** Admissible and inadmissible transpositions

**Step 3** For each decomposition obtained in Step 2, compute the corresponding total optimal cost by summing the optimal cost for all the transpositions in the decomposition.

**Step 4** Determine the optimal product of transpositions for formation reconfiguration by minimizing the total cost over all admissible products of transpositions.

As mentioned earlier, collision between a pair of observer-spacecraft during transposition may occur without collision-avoidance maneuvers. Therefore, we propose to to move small groups of spacecraft in sequence, each group containing more than two spacecraft, to achieve formation reconfiguration. To facilitate the subsequent development, we introduce a few definitions.

**Definition 6.1** Let $\sigma$ and $\sigma'$ be two permutations of $\mathcal{I}$. Then $\sigma$ and $\sigma'$ are *disjoint* if every integer in $\mathcal{I}$ moved by $\sigma$ is fixed by $\sigma'$, and every integer moved by $\sigma'$ is fixed by $\sigma$.

**Definition 6.2** A permutation $\sigma$ of $\mathcal{I}$ is a *cycle of length $m$* or *m-cycle*, if it is a permutation of a subset $\{s'_1, \ldots, s'_m\}$ of $\mathcal{I}$ that replaces $s'_1$ by $s'_2$, $s'_2$ by $s'_3, \ldots, s'_{m-1}$ by $s'_m$, and $s'_m$ by $s'_1$.

Evidently, two disjoint permutations $\sigma$ and $\sigma'$ commute, i.e. $\sigma\sigma' = \sigma'\sigma$. Also, a transposition is a cycle of length 2. We know that given any initial and desired formations corresponding to permutations of $\mathcal{I}$ represented by $s_o$ and $s_d$ respectively, there exists a permutation matrix $\mathbb{P}_{(s_d, s_o)}$ which is decomposable into a product of transpositions. Now, the main question is whether it is possible to decompose $\mathbb{P}_{(s_d, s_o)}$ into a product of *permutation cycles of length $m > 2$*. To answer the foregoing question, we make use of the following basic property of permutation groups [19]:

**Property B** *Let $\sigma$ be a permutation of $\mathcal{I}$. Then $\sigma$ can be expressed as a product of disjoint cycles. This cycle decomposition is unique up to re-arrangement of the cycles involved.*

From Definitions 6.1 and 6.2, disjoint cycles imply that no two cycles move a common point. Note that the cycles in the decomposition may have lengths $\geq 2$. Property B suggests that we should first seek the decompositions of the given $\mathbb{P}_{(s_d, s_o)}$ in terms of disjoint cycles. If all the disjoint cycles have lengths $> 2$, then we can reconfigure the formation by a composition of permutations of small spacecraft groups containing more than two spacecraft. In the actual formation reconfiguration process, all the spacecraft associated with any cycle must move simultaneously. But the disjoint cycles can be initiated at different times. The foregoing approach can be illustrated by the following example.

*Example 6.2* Consider four point-observers in diamond-shaped cluster-formation patterns. Here, the index set $\mathcal{I}$ is $\{1, 2, 3, 4\}$. Let the initial formation (see Fig. 6.13) be represented by $\mathbf{s}_o = [1, 2, 3, 4]^T$. The index set $\mathcal{I}$ has 4! permutations including the identity permutation. These permutations can be classified into the following categories: Six cycles of length 4:

$$\mathcal{C}_4 = \{[4, 1, 2, 3]^T, [2, 3, 4, 1]^T, [2, 4, 1, 3]^T, [3, 1, 4, 2]^T, [3, 4, 2, 1]^T, [4, 3, 1, 2]^T\};$$
$$(6.104)$$

Eight cycles of length 3:

$$\mathcal{C}_3 = \{[4, 1, 3, 2]^T, [2, 4, 3, 1]^T, [3, 1, 2, 4]^T, [2, 3, 1, 4]^T, [3, 2, 4, 1]^T,$$
$$[4, 2, 1, 3]^T, [1, 3, 4, 2]^T, [1, 4, 2, 3]^T\};$$
$$(6.105)$$

Six cycles of length 2:

$$\mathcal{C}_2 = \{[2, 1, 3, 4]^T, [4, 2, 3, 1]^T, [1, 2, 4, 3]^T, [1, 3, 2, 4]^T, [3, 2, 1, 4]^T, [1, 4, 3, 2]^T\};$$
$$(6.106)$$

and Three non-cycles:

$$\mathcal{N} = \{[2, 1, 4, 3]^T, [4, 3, 2, 1]^T, [3, 4, 1, 2]^T\}.$$
$$(6.107)$$

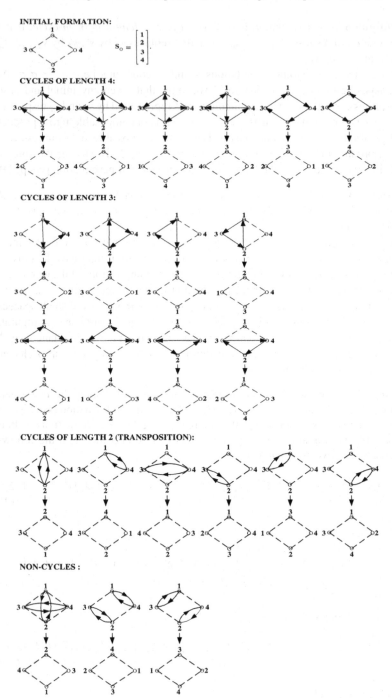

**Fig. 6.13** Decomposition of four-observer cluster-formation into cycles

The formations corresponding to these permutations are illustrated in Fig. 6.13. The spacecraft movements are indicated by arrows. Now, if the desired observer-spacecraft cluster-formation $s_d$ belongs to $C_4$ or $C_3$, then formation reconfiguration can be achieved by a single cyclic permutation without transposition. The total optimal cost can be computed by summing the optimal cost associated with the optimal basic maneuvers of observer-spacecraft in the permutation cycle. If $s_d$ belongs to $C_2$ or $\mathcal{N}$, then formation reconfiguration can be obtained by transpositions or product of cycles of lengths $> 2$. For example, the desired formation $s_d = [3, 4, 1, 2]^T \in \mathcal{N}$ can be attained from the initial formation $s_o = [1, 2, 3, 4]^T$ by two disjoint transpositions: $s_o \rightarrow [3, 2, 1, 4]^T \rightarrow s_d$, or by two successive nondisjoint cycles of length 3: $s_o \rightarrow [3, 2, 4, 1]^T \rightarrow s_d$. We note from Fig. 6.13 that for the first four cycles of length 4, collision between spacecraft occurs when the paths connecting the 1, 2 and 3, 4 positions in the formation cross each other at the same time. However, collision may be avoided by making the transition times associated with the paths connecting positions 1, 2 and 3, 4 different from each other.

**(ii) Type 2 Formation Reconfiguration**
In this type of cluster-formation reconfiguration, a specified position in the desired formation may be occupied by *any* spacecraft of a particular type. As before, the initial and desired formation patterns at time $t$ are specified by the point sets $C^o(t) = \{r_{(1)}^o(t), \ldots, r_{(N)}^o(t)\}$ and $\mathcal{P}^d(t) = \{p_{(1)}^d(t), \ldots, p_{(N)}^d(t)\}$ respectively. We assume that each element $r_{(j)}^d(t) \in C^d(t)$ corresponds to a unique $p_{(i)}^d(t)$ for some $i \in \mathcal{I} = \{1, \ldots, N\}$.

First, we consider the simplest case where all $N$ spacecraft are identical. Moreover, $r_{(i)}^o(t)$ and $r_{(i)}^d(t) = r_j^d(t)$ are related by a translation given by (6.96). In this case, there are $N!$ permutations for formation reconfiguration. However some permutations may not be admissible due to the fact that $r_{(i)}^d(t)$ is not visible from $r_{(i)}^o(t)$ for some $i \in \mathcal{I}$. The solution to the minimum-fuel reconfiguration problem can be solved by first computing the total minimum-fuel expenditure associated with each admissible permutation, and then determine the permutation with the least total minimum-fuel expenditure.

Next, we consider the general case where $r_{(i)}^d(t)$ and $r_{(i)}^o(t)$ have the form:

$$r_{(i)}^d(t) = r_{(i)}^d(0) + \xi(t), \qquad r_{(i)}^o(t) = r_{(i)}^o(0) + \xi(t), \quad i = 1, \ldots, N, \qquad (6.108)$$

where $\xi(t)$ is a specified formation drift vector. Thus, the initial and desired formation patterns are shape-invariant over some time interval $I_{t_f}$. Let $C^o(t)$ (resp. $C^d(t)$) be partitioned into $P$ disjoint subsets of identical spacecraft represented by index set $S_j^o$ (Resp. $S_j^d$) $\subset \mathcal{I}$ with $\cup_{j=1}^P S_j^d = \cup_{j=1}^P S_{j=1}^o = \mathcal{I}$. We assume that $S_j^o$ and $S_j^d$ have the same number of elements $n_j$. Since any element of $S_j^d$ may be replaced by any element of $S_j$, therefore we have a total of $N! / \Pi_{j=1}^P (n_j!)$ distinct formations. Again, we use the column vector $\mathbf{s}_j = (s_{j1}, \ldots, s_{jn_j}^T)$ with $s_{ji} \in S_j$, and $s_{ji} = s_{ji'}$ if and only if $i = i'$, to denote a permutation of $S_j$. Thus, the initial and desired formations can be represented by $s_o = [s_{o1}^T, \ldots, s_{oP}^T]^T$ and $s_d = (s_{d1}^T, \ldots, s_{dP}^T]^T$

respectively. Given $s_o$ and $s_d$, the permutation matrix $\mathbb{P}_{(s_d,s_o)}$ relating $s_o$ and $s_d$ may not be unique, since there may exist many ways for attaining $s_{di}$ by the components of $s_{oi}$. For a given $\mathbb{P}_{(s_d,s_o)}$, we may seek its decomposition as a product of disjoint cycles as proposed earlier for Type 1 formation reconfiguration, and determine the minimum-fuel decomposition corresponding to $\mathbb{P}_{(s_d,s_o)}$. Then the solution to the minimum-fuel formation reconfiguration problem can be found by considering the minimum-fuel admissible decompositions for all possible permutations associated with $\mathbb{P}_{(s_d,s_o)}$.

To obtain a practical solution to the foregoing problem, we propose to simplify the problem by first identifying those spacecraft in the initial formation that match both the positions and spacecraft types in the desired formation. These spacecraft will remain *fixed* relative to the formation during reconfiguration. Then, we seek a solution to the minimum-fuel reconfiguration problem for the *remaining* spacecraft using the method described earlier. This simplified approach may lead to a sub-optimal solution when the fuel expenditures for moving the fixed spacecraft are small relative to those for moving the remaining spacecraft. The results of a computer simulation study for a typical observer-cluster are presented in [20].

### 6.2.1.5 Visibility-Based Optimal Cluster-Formation Motion Planning

Now, assume that the observer-spacecraft are in the desired triangular formation, we proceed to consider the cluster-formation motion planning problem by regarding the observer-spacecraft cluster as a single rigid body with concentrated mass $M = \sum_{i=1}^{3} M_i$ located at the mass center of the cluster. The observer-cluster motion planning problem may be formulated in many ways depending on the mode of interferometer operation that calls for certain type of orbit or trajectory for observation. For example, in the star-acquisition mode, the combiner may be equipped with a wide-aperture camera on-board for capturing the star-light, one may define the optimal motion to be in the direction towards the region of maximum luminosity. To minimize energy expenditure during the searching maneuver, minimum-fuel controls may be used.

In other applications such as using an observer-spacecraft cluster to observe a planetary surface or a three-dimensional object such as an asteroid, it is necessary to consider the visible set of the observer-cluster with respect to the object $\mathcal{O}$ defined by

$$\mathcal{V}(\mathcal{C}) \overset{\text{def}}{=} \bigcup_{i=1}^{K} \mathcal{V}(\Pi_{\mathcal{W}} s_i(t; u_{(i)}),  \qquad (6.109)$$

under the assumption that visibility-blocking does not exist among the observer-spacecraft (see Fig. 6.14 for illustration), and the motions of the observer-spacecraft are describable by equations in the form (6.2). One may formulate the optimal cluster-formation motion planning problem as one of maximizing the integrated total measure of visible set of the observer-cluster (i.e., Problem 6.1).

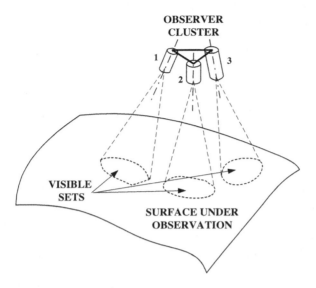

**Fig. 6.14** Visible set of an observer-cluster

## 6.3 Non-cooperative Observation

In many real-world situations, the observers compete with each other to acquire the most observation data for $\mathcal{O}$ over a given observation time interval. Some of these optimal motion planning problems can be formulated in the framework of differential games [21–23]. We shall consider physically meaningful problems by first examining a simple specific example.

In the simplest case, each point-observer plans its motion to maximize its visibility functional over a given time-interval without considering the state and actions of other observers. The corresponding optimal motion planning problem can be stated as follows:

**Problem 6.3** Let $I_{t_f} = [0, t_f]$ be a specified finite observation time-interval, and $s_{(i)}(0) = x_{(i)o}, i = 1, 2$ be given initial states of the point-observers at $t = 0$. For each $i = 1, 2$, find an admissible control $u_{(i)}(\cdot) \in \mathcal{U}_{ad}^i(I_{t_f})$ defined on $I_{t_f}$ such that its corresponding motion $s_{(i)}(t; u_{(i)}) \in \mathcal{P}_i$ for all $t \in I_{t_f}$, and the visibility functional

$$J_i(u_{(i)}) \overset{\text{def}}{=} \int_0^{t_f} \mu_n\{\mathcal{V}(\Pi_{\mathcal{W}} s_{(i)}(t; u_{(i)}))\} dt \tag{6.110}$$

takes on its maximum value.

Another problem involves two observers. Observer 1 tries to obtain as much visual data on object $\mathcal{O}$ as possible over the given observation time-interval. Observer 2 has an opaque solid body which can be used to block the view of Observer 1 thereby reducing its visibility of object $\mathcal{O}$ (See Fig. 6.15). This problem can be posed as follows:

**Problem 6.4** Given a finite observation time interval $I_{t_f} = [0, t_f]$, and the initial states of the observers at $t = 0$, $s_{(i)}(0) = s_{(i)o}$, $i = 1, 2$, find admissible controls $u_{(i)}(\cdot) \in \mathcal{U}_{ad}^{(i)}(I_{t_f})$, $i = 1, 2$ defined on $I_{t_f}$ such that

$$\sup_{u_{(2)} \in \mathcal{U}_{ad(2)}(I_{t_f})} \inf_{u_{(1)} \in \mathcal{U}_{ad(1)}(I_{t_f})} J_\lambda(u_{(1)}, u_{(2)}) \tag{6.111}$$

is attained, where $J_\lambda$ is the weighted visibility functional defined by (6.5).

Before we proceed to develop optimality conditions for Problems 6.3 and 6.4, we shall make a few intuitive observations on the nature of solutions to these problems with the aid of the special case where the object $\mathcal{O}$ under observation is a smooth surface described by the graph of a $C_1$-function $f = f(x_1, x_2)$ defined on a compact set $\Omega$ in $\mathbb{R}^2$.

First, consider Problem 6.3. Obviously, this problem corresponds to two independent optimization problems. We recall from Proposition 3.1 in Chap. 3 and its extension to the case with $\dim(\Omega) = 2$, there exists a critical vertical-height surface $h_{vc} = h_{vc}(x_1, x_2)$ such that $\mathcal{O}$ is totally visible from any point on this surface. Evidently, for Observers 1 and 2 to attain the highest values of $J_1$ and $J_2$ respectively at the terminal time $t_f$, they must try to reach the critical vertical-height surface as quickly as possible and remain there as long as possible within the given observation time-interval $I_{t_f}$. Note that the point-observers in this problem have no visibility interference between them.

Another necessary condition for optimality can be obtained by introducing a new state variable $y$. The evolution of $y(t)$ with time $t$ is described by

$$\dot{y}(t) = \mu_n\{\mathcal{V}(\Pi_{\mathcal{W}}s_u(t))\}, \quad y(0) = 0. \tag{6.112}$$

Thus, $J_1(u) = y(t_f)$. Let the Hamiltonian associated with the combined system (6.2) and (6.11) be defined by:

$$\mathcal{H}(s, \eta, u) = \mu_n\{\mathcal{V}(\Pi_{\mathcal{W}}s)\} + \eta^T[\varphi(s) + \chi(s)u], \tag{6.113}$$

where $\eta = [\eta_1, \ldots, \eta_N]^T$, $N = n_1 + n_2 + 1$, corresponds to the state of the adjoint system:

$$\dot{\eta} = -\nabla_s \mathcal{H} = -\nabla_s \mu_n\{\tilde{\mathcal{V}}(\Pi_{\mathcal{W}}s)\} - (D_s\varphi)^T \eta - \sum_{i=1}^{n} \eta_i \left( \sum_{j=1}^{m} \nabla_s \chi_{ij}(s)u_j \right). \tag{6.114}$$

Again, if the function $s \to \mu_n\{\mathcal{V}(\Pi_{\mathcal{W}}s)\}$ on $\mathbb{R}^n \to \mathbb{R}^+$ is smooth, then a necessary condition for optimality is given by:

**Theorem 6.3** *Suppose that the function $s \rightarrow \mu_n\{\mathcal{V}(\Pi_{\mathcal{W}}s)\}$ on $\mathbb{R}^n \rightarrow \mathbb{R}^+$ is $C_1$. Let $u^* = u^*(t)$ be an optimal control for Problem 6.2 with corresponding response $s^* = s^*(t)$. Then there exists an absolutely continuous function $\eta^* = \eta^*(t)$ satisfying (6.114) for almost all $t \in I_{t_f}$, where*

$$\mathcal{H}(s^*(t), \eta^*(t), u^*(t)) = \mathcal{M}(s^*(t), \eta^*(t)) \overset{\text{def}}{=} \max_{u \in U} \mathcal{H}(s^*(t), \eta^*(t), u). \quad (6.115)$$

*Equations (6.2), (6.13) and (6.14) with initial and terminal conditions:*

$$(s, y)(0) = (s_o, 0), \quad \eta(t_f) = 0 \quad (6.116)$$

*constitute a nonlinear two-point-boundary-value problem for which the optimal motion $(s^*, y^*)(\cdot)$ must satisfy.*

In Problem 6.4, only Observer 2 has an opaque spherical solid-body, and it tries to block the visibility of Observer 1 for observing $\mathcal{O}$. To accomplish this task, Observer 2 attempts to enter as quickly as possible Observer 1's visibility cone $\mathcal{C}$ with vertex at $x_{(1)}(t; u_{(1)})$, containing the object $\mathcal{O}$. In the critical case for complete blockage of the visibility of $\mathcal{O}$ from Observer 1, the spherical body of Observer 2 is tangent to the cone $\mathcal{C}$ as illustrated in Fig. 6.15. To minimize the visible data on the object $\mathcal{O}$ acquired by Observer 1 over a given finite observation time interval, Observer 2 should try to move in such a way to maintain maximum blockage of the visibility of $\mathcal{O}$ from Observer 1 as long as possible.

To provide further insight into the solutions of the aforementioned non-cooperative optimal motion planning problems, we consider a simple specific example.

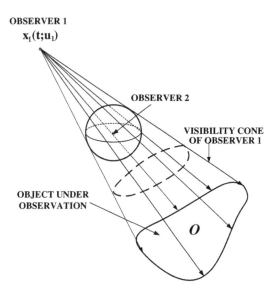

**Fig. 6.15** Critical case for complete blockage of the visibility of $\mathcal{O}$ from *Observer 2* by *Observer 1*

OBSERVER 1
$x_1(t;u_1)$

OBSERVER 2

VISIBILITY CONE OF OBSERVER 1

OBJECT UNDER OBSERVATION

$\mathcal{O}$

*Example 6.3* Let the object under observation be a stationary opaque circular disk with radius $R_o$, centered at the origin of the world space $\mathcal{W} = \mathbb{R}^2$. Observer 1 is composed of an opaque massless opaque circular disk with radius $R_1$, and a concentrated mass $M_1$ at the disk center which is constrained to move along a circle with radius $r_1 > R_o + R_1$, centered at the origin. Observer 2 is a point-observer with mass $M_2$ whose motion is confined to a concentric circle with radius $r_2 > r_1 + R_1$. For simplicity, let the angular positions $\theta_i$ of the $i$th Observer be described by Newton's law:

$$M_i r_i \ddot{\theta}_i = u_{(i)}, \quad i = 1, 2, \tag{6.117}$$

where $u_{(i)}$ is the tangential force of the $i$th observer. We assume there exist radial forces that keep the observers in the specified circular paths at all times. Moreover, the controls $u_{(i)}, i = 1, 2$ are magnitude limited and take their values in the compact sets:

$$U_{(i)} = \{u_{(i)} \in \mathbb{R} : |u_{(i)}| \leq F_i\}, \quad i = 1, 2, \tag{6.118}$$

where $F_i$'s are specified positive constants.

Now, given the observers' angular positions $(\tilde{\theta}_1, \tilde{\theta}_2)$, the corresponding visible set of Observer 2 can be determined geometrically. Figure 6.16 shows three possible cases. Figure 6.16a corresponds to the case where there is no visual blockage of Observer 2 by Observer 1. in this case, the visible set of Observer 2 at $(r_2, \tilde{\theta}_2)$ in the presence of Observer 1 at $(r_1, \tilde{\theta}_1)$ is a circular arc given by

$$\mathcal{V}_a((r_2, \tilde{\theta}_2)|(r_1, \tilde{\theta}_1)) = \{(r, \theta) : r = R_o, \tilde{\theta} \tag{6.119}$$

Figure 6.12b corresponds to the case where there is partial visual blockage of Observer 2 by Observer 1. The visible set of Observer 2 in the presence of Observer 1 for this case is a connected proper subset of $\mathcal{V}_a((r_2, \tilde{\theta}_2)|(r_1, \tilde{\theta}_1))$. Finally, Fig. 6.16c also shows the case where there is partial visual blockage of Observer 2 by Observer 1. However, in this case, the visible set of Observer 2 in the presence of Observer 1 is the union of two disconnected proper subsets of $\mathcal{V}_a((r_2, \tilde{\theta}_2)|(r_1, \tilde{\theta}_1))$.

Now, since maximum visual blockage of Observer 2 by Observer 1 is attained when $\theta_1$ is equal to $\theta_2$. Thus, Observer 1 tries to align its angular position $\theta_1$ with $\theta_2$. On the other hand, to avoid visual blockage of Observer 2 by Observer 1, Observer 2 tries to avoid alignment of its angular position $\theta_2$ with $\theta_1$. For both observers, it is natural to consider the angular positional deviation $\theta_1 - \theta_2$. In view of the control magnitude limitations (6.17), fastest alignment of $\theta_2$ with $\theta_1$, and misalignment of $\theta_1$ with $\theta_2$ can be obtained by applying "bang-bang" controls of the form:

$$u_{(1)} = -F_1 \text{sgn}(\sigma_1(\theta_2 - \theta_1, \dot{\theta}_2 - \dot{\theta}_1)), \quad u_{(2)} = F_2 \text{sgn}(\sigma_2(\theta_1 - \theta_2, \dot{\theta}_1 - \dot{\theta}_2)), \tag{6.120}$$

where sgn denotes the signum function:

**Fig. 6.16** Two-observer system in Example 6.1, **a** no visual blockage by observer 1, **b** partial visual blockage by observer 1, **c** partial visual blockage by observer 1

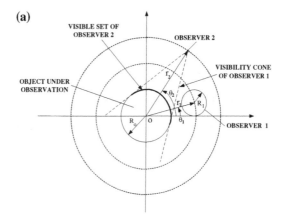

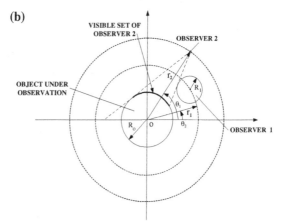

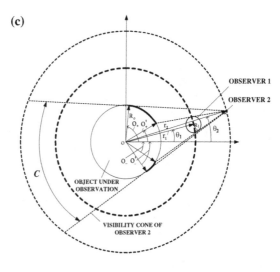

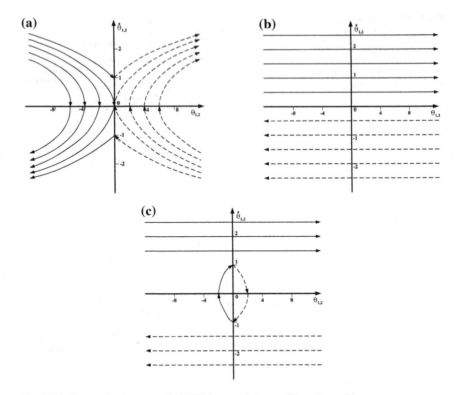

**Fig. 6.17** Sketch of trajectories of (6.122) for case (**a**), case (**b**), and case (**c**)

$$\text{sgn}(q) = \begin{cases} +1, & \text{if } q > 0, \\ 0 & \text{if } q = 0, \\ -1 & \text{if } q < 0, \end{cases} \tag{6.121}$$

and $\sigma_i$, $i = 1, 2$ are real-valued switching functions of their arguments. A typical switching function is given by $\sigma_i(\psi, \dot{\psi}) = \kappa_{1(i)}\psi + \kappa_{2(i)}\dot{\psi}$, where $\kappa_{1(i)}$ and $\kappa_{2(i)}$ are specified real numbers. Equation (6.120) corresponds to feedback controls or strategies for the observers. For the simplest switching function $\sigma_i(\psi, \dot{\psi}) = \psi$, (6.117) and (6.120) lead to a single equation for $\theta_{1,2} \overset{\text{def}}{=} \theta_1 - \theta_2$:

$$\ddot{\theta}_{1,2} = F_{1,2}\text{sgn}(\theta_{1,2}), \tag{6.122}$$

where $F_{1,2} \overset{\text{def}}{=} F_1/(M_1 r_1) - F_2/(M_2 r_2)$. Figure 6.17 shows the trajectories of (6.122) for (a) $F_1/(M_1 r_1) > F_2/(M_2 r_2)$ (b) $F_1/(M_1 r_1) = F_2/(M_2 r_2)$ and (c) $F_1/(M_1 r_1) < F_2/(M_2 r_2)$. In Case (a), the effective control force of Observer 1 exceeds that of Observer 2. Figure 6.17a shows there exist two trajectories passing through the origin of the $(\theta_{1,2}, \dot{\theta}_{1,2})$-plane, implying that from any point $(\theta_{1,2}(0), \dot{\theta}_{1,2}(0))$ on this trajectory (i.e. $\theta_{1,2}(0) = \dot{\theta}_{1,2}(0)^2/(2F_{1,2})$), maximum visual blockage of Observer

2 by Observer 1 is attainable in finite time $t_1 = \dot{\theta}_{1,2}(0)(\sqrt{2} - 1)/F_{1,2}$. However, any perturbation in the relative position and/or velocity will result in drifting away from maximum blockage. In Case (b), the effective control forces of Observers 1 and 2 are equal. The trajectories cross the $\dot{\theta}_{1,2}$-axis, implying that along a trajectory, maximum visual blockage of Observer 2 by Observer 1 can occur at a time-instant only, and then the observers drift away from each other at constant angular velocity as shown in Fig. 6.17b. In Case (c), the effective control force of Observer 1 is less than that of Observer 2. Thus, Observer 2 is capable of avoiding alignment of its angular position $\theta_2$ with $\theta_1$ (see Fig. 6.17c). At this point, we note that for Observer 1 to implement the foregoing control strategy to achieve maximum blockage of Observer 2, it is necessary for Observer 1 to have complete knowledge of the state of Observer 2 at any time. This requirement implies that Observer 1 must have appropriate sensors for performing this task. Now, if Observer 2 is capable of detecting the presence of Observer 1 in blocking the view of the object, it could try to reduce the blocking if it has complete knowledge of the state of Observer 1.

The foregoing feedback controls for Example 6.1 are obtained by intuitive reasoning. Now, we shall proceed to develop necessary conditions for which the optimal controls must satisfy formally.

Consider Problem 6.4 for non-cooperative observers whose motions are described by (6.1). Let $u^*(\cdot) = (u^*_{(1)}(\cdot), u^*_{(2)}(\cdot))$ be an optimal control pair with corresponding motions $x_u^*(\cdot)$. It follows that for $u_{(1)} = u^*_{(1)}$, $J_\lambda(u^*_{(1)}, u_{(2)})$ cannot increase for any admissible perturbed $u_{(2)}$. Also, for $u_{(2)} = u^*_{(2)}$, $J_\lambda(u_{(1)}, u^*_{(2)})$ cannot decrease for any admissible $u_{(1)}$. Now, if the function $x \to \mu_n\{\mathcal{V}(x)\}$ on $\Omega \to \mathbb{R}^+$ is $C_1$, then $DJ_\lambda(u^*, \delta u_{(2)}) \le 0$ for all admissible $u_{(2)}$, and $DJ_\lambda(u^*, \delta u_{(1)}) \ge 0$ for all admissible $u_{(1)}$. Moreover,

$$\sup_{u_{(2)} \in \mathcal{U}^{(2)}_{ad}(I_{t_f})} \inf_{u_{(1)} \in \mathcal{U}^{(1)}_{ad}(I_{t_f})} J_\lambda(u_{(1)}, u_{(2)}) = \inf_{u_{(1)} \in \mathcal{U}^{(1)}_{ad}(I_{t_f})} \sup_{u_{(2)} \in \mathcal{U}^{(2)}_{ad}(I_{t_f})} J_\lambda(u_{(1)}, u_{(2)}; \lambda). \quad (6.123)$$

Thus, the optimal control $u^*$ is a saddle point of the weighted visibility functional $J_\lambda$ defined on the $(u_{(1)}, u_{(2)})$-space. We may summarize the foregoing conclusion as a necessary condition for which an optimal control must satisfy (illustrated by Fig. 6.14):

**Theorem 6.4** *Suppose that an optimal control* $u^* = (u^*_{(1)}, u^*_{(2)})(t)$ *defined on* $I_{t_f}$ *for Problem 6.2 involving* $J_\lambda$ *defined by (6.5) exists, and the function* $s \to \mu_n\{\mathcal{V}(s)\}$ *on* $\Omega \to \mathbb{R}^+$ *is* $C_1$*. Then* $u^*(\cdot)$ *must satisfy the following pair of variational inequalities:*

$$DJ_\lambda(u^*, \delta u_{(2)}) = \int_0^{t_f} \lim_{\alpha \to 0^+} \frac{1}{\alpha} (\lambda \mu_n\{\mathcal{V}(\Pi_W s_{(1)}(t; u^*_{(1)})|\Pi_W s_{(2)}(t; u^*_{(2)} + \alpha \delta u_{(2)}))\}$$
$$+ (1 - \lambda) \mu_n\{\mathcal{V}(\Pi_W s_{(2)}(t; u^*_{(2)} + \alpha \delta u_{(2)})|\Pi_W s_{(1)}(t; u^*_{(1)}))\}) dt \le 0 \quad (6.124)$$

*for all admissible* $u^*_{(2)} + \alpha \delta u_{(2)}, 0 \le \alpha < 1,$ *and*

**Fig. 6.18** Display of the
basic features associated with
the saddle-like structure of
$J_\lambda$ in the neighborhood of
$(u^*_{(1)}, u^*_{(2)})$ as described in
Theorem 6.4

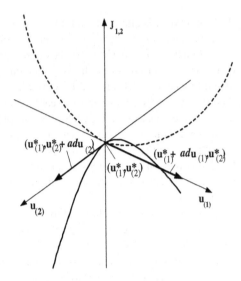

$$DJ_\lambda(u^*, \delta u_{(1)}; \lambda) = \int_0^{t_f} \lim_{\alpha \to 0^+} \frac{1}{\alpha}(\lambda\mu_n\{\mathcal{V}(\Pi_\mathcal{W}s_{(1)}(t; u^*_{(1)} + \alpha\delta u_{(1)})|\Pi_\mathcal{W}s_{(2)}(t; u^*_{(2)}))\}$$
$$+ (1 - \lambda)\mu_n\{\mathcal{V}(\Pi_\mathcal{W}s_{(2)}(t; u^*_{(2)})|\Pi_\mathcal{W}s_{(1)}(t; u^*_{(1)} + \alpha\delta u_{(1)}))\})dt \geq 0$$

$$(6.125)$$

*for all admissible* $u^*_{(1)} + \alpha\delta u_{(1)}$, $0 \leq \alpha < 1$. *Moreover, condition (6.123) is satisfied*
*(Fig. 6.18).*

## 6.4 Problems for Further Study

### 6.4.1 Observation of Objects with Time-Varying Shapes

So far, we have only considered the problem of optimal motion planning for observing
objects with time-invariant shapes. When the geometric shapes of observed objects
vary with time, the complexity of the optimal observer-motion planning problem
depends on the available information on the shape dynamics of the observed objects.
In practical applications, it is of interest to plan the observer motion in real-time based
on the available observation data. If we have a model for the shape-dynamics of the
observed objects, then we can use this model to predict the objects' shapes over a
future time interval, and use the predicted object shapes to plan the observer motion.
To achieve complete visibility of the objects, it may be necessary to use multiple
observers to perform the task. In this case, motion planning will involve developing
real-time decision rules for dividing the observation task among the observers.

## 6.4.2 Cooperative Observation Involving Multiple Point-Observers with Communication Time-Delays

In the cooperative observation of objects using multiple point-observers, it is desirable to have observers communicating with each other to minimize redundant observation and to maintain their relative positions and attitudes in the presence of external disturbances. In real communication systems, time-delays in data transmission and processing are unavoidable. Such time-delays may have detrimental effects on the performance of the observation process and destabilization of relative position and attitude control systems. Therefore, a detailed study of such effects is necessary before physical implementation of the multi-observer system.

## 6.4.3 Generalized Visibility-Based Optimal Path and Motion Planning Problems

In this monograph, we have focused our attention primarily on optimal path and motion problems involving the measures of visible sets of the observers. In many physical situations, one may be interested in certain special features or properties of the objects under observation. For example, in planetary exploration, the planetary surface features such as color and texture are related to the chemical composition of the surface. The presence of green color on a planetary surface may suggest the presence of plant life. In the formulation of optimal path and motion planning problems, one may replace the surface measure of the visible set $\mu_n\{\mathcal{V}(s)\}$ by another real-valued function of $\mathcal{V}(s)$ derived from certain special properties of the observed objects. Evidently, studies in this direction should be closely tied to the physical situation.

## Exercises

**Ex. 6.1** This exercise can be regarded as an extension of Problem 5.1 in Chap. 5. Here, we consider *two* point-observers whose motions are confined to a hemispherical platform in $\mathbb{R}^3$ with radius $r_p$ as shown in Fig. 6.19. The object under observation is a hemisphere with radius $r_o < r_p$. Let the mass of the $i$th point-observer be denoted by $M_i$.

(a) Assuming the absence of friction forces, write down the equations of motion for the point-observers in spherical coordinates, and the necessary radial forces to keep the point-observers on the observation platform. Also, the $i$th point-observer has a control force $u_{(i)}$.

**Fig. 6.19** Sketch of two point-observers moving on a hemispherical platform (Ex. 6.1)

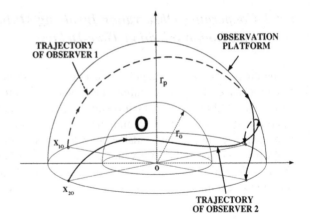

(b) Give a mathematical formulation of the problem for optimal motion-planning for non-cooperative observation of the hemisphere over a given finite observation time-interval.

**Ex. 6.2** Using the geometric relations between the object and the observers, determine the visible sets of Observer 2 for the cases shown in Fig. 6.3b, c in terms of the parameters $\theta_1, \theta_2, R_o, R_1, r_1$ and $r_2$ satisfying conditions $r_1 > R_o + R_1$ and $r_2 > r_1 + R_1$.

**Ex. 6.3** Consider the question posed in Remark 6.3. namely, for two observers with solid bodies modelled by opaque closed spherical balls $\bar{\mathcal{B}}(\Pi_{\mathcal{W}}s_{(i)}(t; u_{(i)}); r_i)$, $i = 1, 2$, centered at $\Pi_{\mathcal{W}}s_{(i)}(t; u_{(i)})$ with finite radius $r_i$, does blocking reduction of visibility always increase the value of $J_\lambda$ defined in Eq. (6.7) If your answer is "true"or "false", give a proof. If your answer is "sometimes true and sometimes false", give a simple example for each case to support your answer.

**Ex. 6.4** Propose an algorithm for computing the visible set of an observer with finite viewing aperture-angle for observing a 2D-surface in $\mathbb{R}^3$ in the presence of more than one visibility-blocking observers modelled by opaque solid spheres as illustrated in Fig. 6.1 (with only one visibility-blocking observer).

**Ex. 6.5** Consider the formation reconfiguration of an observer-cluster consisting of two sets of different types of observers (represented by black and white disks) whose initial and desired formation patterns correspond to eight equally-spaced points on a circle as illustrated by Fig. 6.20.

(a) Determine the number of distinct cluster-formation patterns that can be generated by permutating the observers' positions?
(b) Assume that only Observers 1, 4, 5 and 6 are allowed to move relative to the cluster, determine all their permutation cycle of length $>2$ for attaining the desired cluster-formation pattern.

**Fig. 6.20** Initial and desired formation patterns for an eight observer-cluster described in Ex. 6.5

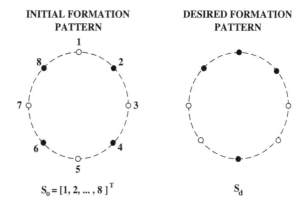

INITIAL FORMATION
PATTERN

$S_0 = [1, 2, \dots, 8]^T$

DESIRED FORMATION
PATTERN

$S_d$

(c) Assume that only Observers 1, 2, 4, 5, 6 and 8 are allowed to move while keeping Observers 3 and 7 fixed relative to the cluster-formation, determine all products of two disjoint cycles of length 3 for attaining the desired cluster-formation pattern.

(d) Use the results on the basic minimum-fuel formation reconfiguration maneuvers discussed in Sect. 6.2.1.4, obtain solutions for the minimum-fuel formation reconfiguration problem associated with part (c).

# References

1. P.K.C. Wang, Navigation strategies for multiple mobile robots moving in formation. J. Rob. Syst. **8**(2), 177–195 (1991)
2. P.K.C. Wang, F.Y. Hadaegh, Coordination and bcontrol of multiple microspacecraft moving in formation. J. Astronaut. Sci. **44**(3), 315–355 (1996)
3. P.K.C. Wang and F.Y. Hadaegh, Simple Formation-keeping Control Laws for Multiple Microspacecraft, UCLA Engr. Rpt. 95–130, 1995
4. P.K.C. Wang, F.Y. Hadaegh, K. Lau, Synchronized formation rotation and attitude control of multiple free-flying spacecraft. J. Guidance Control Dyn. **24**(2), 352–359 (2001)
5. M.D. Johnson, K.T. Nock, Multiple Spacecraft optical interferometry trajectory analysis, Workshop on technologies for space interferometry, Jet Propulsion Laboratory, Pasadena, Calif. 30 April–2 May 1990
6. R. Stachnik, K. Ashlin, S. Hamilton, Space station-SAMSI: a spacecraft array for michelson spatial interferometry. Bull. Am. Astron. Soc. **16**(3), 818–827 (1984)
7. R.V. Stachnik et al., Multiple spacecraft Michelson stellar interferometry. Proc. SPIE, Instrum. Astron. **445**, 358–369 (1984)
8. W. Hahn, *Stability of Motion* (Springer, New York, 1967), p. 275
9. Formation Flying of Multiple Spacecraft, JPL Video No.AVC-97-039, Jet Propulsion Lab. Pasadena, 5 1996
10. J. Credland et al, Cluster - ESA's spacefleet to the magnetosphere, ESA Bull. No.84, 1995
11. A.B. DeCue, Multiple spacecraft optical interferometry, Preliminary feasibility assessment, JPL Technical Internal Report D-8811, 1991

12. K. Lau, M. Colavita, M. Shao, *The New Millennium Separated Spacecraft Interferometer,* Presented at the Space Technology and Applications International Forum (STAIF-97) (Albuquerque, NM, 1997), pp. 26–30
13. D.F. Lawden, *Optimal Trajectories for Space Navigation* (Butterworth, London, 1963)
14. J.P. Marec, Optimal transfer between close elliptic orbits, NASA Tech. Translation, F-554, 1969
15. J.P. Marec, *Optimal Space Trajectories* (Elsevier Scientific Publication, New York, 1979)
16. T.E. Carter, Fuel-optimal maneuvers of a spacecraft relative to a point in circular orbit. J. Guidance **7**(6), 710–716 (1984)
17. D.J. Bell, Optimal space trajectories, a review of published work. Aeronaut. J. **72**, 141–146 (1968)
18. M. Athans, P.L. Falb, *Optimal Control* (McGraw-Hill, New York, 1966)
19. J.F. Humphreys, M.Y. Prest, *Numbers* (Cambridge University Press, Cambridge, Groups and Codes, 1989)
20. P.K.C. Wang, F.Y. Hadaegh, Minimum fuel formation reconfiguration of multiple free-flying spacecraft. J. Astronaut. Sci. **47**(1, 2), 77–102 (1999)
21. A. Friedman, *Differential Games* (Wiley, New York, 1971)
22. R. Isaacs, *Differential Games* (Wiley, New York, 1965)
23. N.N. Krasovskii, A.I. Subbotin, *Game-Theoretic Control Problems* (Springer, New York, 1985)

# Appendix A
# Set Covering Problem

The set covering problem can be stated as follows:

Given a finite set $X$ with $m$ elements, and a family $\mathcal{F}$ of subsets of $X$:

$$\mathcal{F} = \{S_1, \ldots, S_n\}, \quad S_i \subseteq X, \quad i \in \mathcal{I}_n = \{1, \ldots, n\},$$

find a set $\mathcal{I}^* \subseteq \mathcal{I}_n$ with the smallest cardinal number such that

$$\bigcup_{i \in \mathcal{I}^*} S_i = X.$$

The foregoing problem can be formulated as an Integer Programming Problem: Let $A$ be a $m \times n$ incidence matrix with real elements $a_{ij}$ such that

$$a_{ij} = \begin{cases} 1 & \text{if } s_i \in S_j; \\ 0 & \text{if } s_i \in S_j, \end{cases} \quad j \in \mathcal{I}_n.$$

Let $s = (s_1, \ldots, s_n)^T$. Find a $s^*$ in the set

$$\{s : As \geq 1^m, \quad s_j \in \{0, 1\}, \quad j \in \mathcal{I}_n\}$$

such that

$$C(s) = \sum_{j=1}^{n} s_j$$

is minimized, where $1^m$ denotes the $m$-dimensional column vector with unit components.

The foregoing problem can be solved numerically using suitable Mixed Integer Programming software.

© Springer International Publishing Switzerland 2015                    191
P.K.-C. Wang, *Visibility-based Optimal Path and Motion Planning*,
Studies in Computational Intelligence 568, DOI 10.1007/978-3-319-09779-4

# Appendix B
# Art Gallery Problems

The "Art Gallery Problems" were first formulated by Klee (Ref. [10] in Chap. 3):

## B.1 Interior Art Gallery Problem

Given an art gallery described by a polygon $\Pi_N$ in $\mathbb{R}^2$ with $N$ vertices, find the minimum number and locations of point-guards in $\Pi_N$ such that each edge of $\Pi_N$ (the gallery walls) is visible by at least one of the guards.

For this problem, Chvátal gave a solution pertaining to the minimum number of point-guards.

**Chvátal's Theorem (1975):** $\lfloor N/3 \rfloor$ ("integer floor") point-guards are sufficient and sometimes necessary to cover the interior and the edges of a polygon in $\mathbb{R}^2$ with $N$ vertices.

An example with $N = 7$ is shown in Fig. B.1. Applying the above theorem to this case, $\lfloor 7/3 \rfloor = 2$.

## B.2 Exterior Art Gallery Problem

Given a polygon $\Pi_N$ in $\mathbb{R}^2$ with $N$ vertices, find the minimum number and locations of point-guards exterior to $\Pi_N$ such that each edge of $\Pi_N$ is visible by at least one of the guards.

**Theorem of Aggarwal and O'Rourke (1984):** $\lceil N/3 \rceil$ ("integer ceiling") point-guard are sufficient and sometimes necessary to cover the exterior of a polygon in $\mathbb{R}^2$ with $N$ vertices.

Applying the above theorem to the example with $N = 7$ as shown in Fig. B.2 gives $\lceil 7/3 \rceil = 3$.

© Springer International Publishing Switzerland 2015
P.K.-C. Wang, *Visibility-based Optimal Path and Motion Planning*,
Studies in Computational Intelligence 568, DOI 10.1007/978-3-319-09779-4

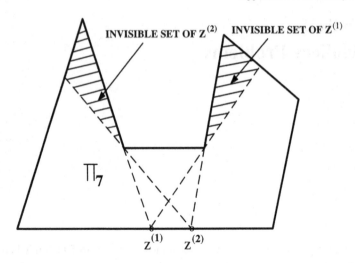

**Fig. B.1** Example of an interior art gallery problem for $\Pi_7$

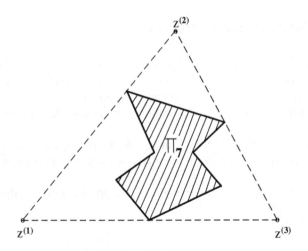

**Fig. B.2** Example of an exterior art gallery problem for $\Pi_7$

# Appendix C
# Delaunay Triangulation

A common method for constructing a geometric figure from a given set of points is to apply first a triangulation algorithm for connecting the points, and then to construct their convex hull. The use of Delaunay triangulation is particularly suitable when we do not wish to impose any constraints on the set of points to be connected. Besides, Delaunay triangulation has some interesting properties such as optimal equiangularity and uniqueness (2D case). The basic idea is to project the points to a paraboloid by summing the squares of their coordinates and taking the convex hull of the projected points. Then the vertical dimension is removed from the lower envelope of the convex hull. The projected edges are the Delaunay triangulation of the original points (see Ref. [1] for further details). Figure C.1 shows a typical mesh obtained by Delaunay triangulation.

© Springer International Publishing Switzerland 2015
P.K.-C. Wang, *Visibility-based Optimal Path and Motion Planning*,
Studies in Computational Intelligence 568, DOI 10.1007/978-3-319-09779-4

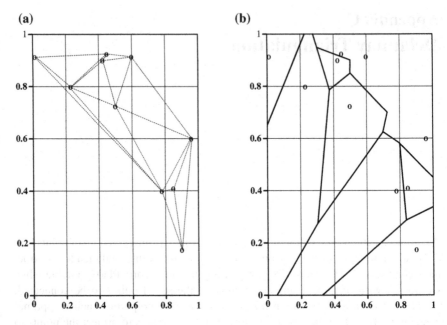

**Fig. C.1 a** A typical mesh obtained by Delaunay triangulation. **b** The corresponding Voronoi diagram

# Reference

1. S. Rippa, Minimal roughness property of the Delaunay triangulation. Comput. Aided Geom. Des. **7**, 489–497 (1990)

# Index

© Springer International Publishing Switzerland 2015
P.K.-C. Wang, *Visibility-based Optimal Path and Motion Planning*,
Studies in Computational Intelligence 568, DOI 10.1007/978-3-319-09779-4

Printed in the United States
By Bookmasters